THE BROADBILL SWORDFISHERY OF THE NORTHWESTERN ATLANTIC

The Fishery's Economic and Natural History
(Early 1800s through 1995)

Charles Dana Gibson

Cover Painting: Richard Ellis
Two makos, probably the only creatures in the ocean fast and powerful enough to attack a broadbill swordfish, close in for the kill.

Ensign Press
PO Box 638
Camden, Maine 04843

Published by Ensign Press
PO Box 638
Camden, Maine 04843

Printed in the United States of America

Library of Congress Catalog Card Number: 97-78416

ISBN: 0-9608996-7-7

THE BROADBILL SWORDFISHERY OF THE NORTHWESTERN ATLANTIC: The Fishery's Economic and Natural History (Early 1800s through 1995)
Gibson, Charles Dana 1928 -
1. Swordfish. 2. Fishery History. 3. Fishing, Commercial. 4. Maritime -- United States

The paper on which this book is printed meets the minimum requirements of American National Standard for Information Science-Permanence of paper for printed library materials. ANSI Z 39.48-1984. Additionally, the paper on which this book is printed meets and exceeds all NASTA specifications.

Main text typeset in Times New Roman 12

TABLE OF CONTENTS

ILLUSTRATIONS AND TABLES

AUTHOR'S NOTE

There are many gaps in the history of the American fisheries. Some of those fisheries no longer economically exist, having become victim to gross over-exploitation. With few exceptions, those that remain are precariously balanced on the slippery slope of survival. The swordfishery of the North Atlantic is one of these that are endangered. It has promise for resource recovery, but only if it is well regulated, an undertaking especially difficult since it is a pelagic fishery whose widespread geographic range will require the involvement of a number of nations. A primary element for the support of any management program must embrace the axiom that *to foretell the future is to know the past.* It is the author's hope that this monograph may be of some help in that regard.

The contents related within this monograph, as they concern people and events, are solely the responsibility of the author and do not in any way reflect the opinions of the publisher.

Charles D. Gibson
Camden, Maine

ACKNOWLEDGMENTS

Part I, **The Fishery,** first appeared during 1981 in an article written by me for *American Neptune* (Vol. XLI, No. 1). It appears again here but with extensive enlargement so as to cover the development of the fishery into the mid 1990s.

I wish to thank those many individuals who I interviewed over the years and who I have accordingly referenced in the text. Many of the interviews took place during the mid-1970s at the time that I was myself engaged in the swordfishery. Further interviews were undertaken during the 1990s. Foremost among those persons who gave of their time and expertise were the following: Martin R. Bartlett who made available to me his personal records which he maintained while working both as a fishery specialist and later as a swordfish skipper; Henry Klimm who, like Bartlett, was not only a pioneer in early longline development but proved to be an especially articulate raconteur of those days. Of the fishery scientists, John J. Hoey, Jack Casey, and Jean Cramer were particularly helpful not only in providing me with various reports but in venturing their personal prognosis for the direction the fishery would take. I am equally indebted to the custodians of the fine technical library which at the time of my early research was housed at the National Marine Fisheries Center at Woods Hole,

MA. It was at that facility that I gathered much of the material involving the early history of the swordfishery in New England.

Part II, **A Corporate Plunge Into Thinning Swordfish Waters: 1964 - 1969** could not have been set down without the generous contribution of Donald Forté, formerly an officer of the Chilmark Fleet Company, who gave me free access, without reservation, to the corporate records, inclusive of the tax returns of the Chilmark Fleet Company; and to Louis Larson who devoted many hours recounting his experience as a skipper with that Company.

Part III, **A Natural History,** brings together the investigative products of the many fishery scientists whom I have listed in my Bibliography. My function, as I viewed it, was to relate this abundance of information within a composite format aimed at a general audience having an interest in the how's and why's of this fascinating fish and its environs.

The cover design is by Ken Gross; the cover painting is by Richard Ellis. The sketches illustrating current systems, statistics, and fishing gear are the products of Winifred Dassman and Ken Gross.

Finally -- and most importantly -- substantial credit goes to my wife, Kay, without whose typing and layout skills this monograph could never have been produced.

CDG

PART I

THE FISHERY

The Beginning Years:
Early Swordfishery in New England

One of the earliest mentionings of swordfish in the northwest Atlantic comes from John Josselyn's *Account of Two Voyages to New England* which was published in London during 1674. Josselyn talked of a great fish called the *vehuella* or *sword-fish* which pierced his ship in what seems to have been an unprovoked attack. The fish broke off its sword and escaped. In 1817, there was a mention in *American Monthly* magazine of a swordfish taken off Sandy Hook by harpoon which when brought into the New York market and sold for food, and despite it being an unknown item, brought a price of a *"quarter of a dollar the pound"*. There is mention in the *Barnstable Patriot* of 30 June 1841 that, *"...fishermen of Martha's Vineyard capture a small amount of swordfish, pickling approximately two hundred pounds a year."* Thus it is established for the record that the Vineyard was an early base for the swordfishery.

Again, according to the *Barnstable Patriot* of 20 August 1861, "*thirty vessels are fifteen miles south and east of No Mans Land or sixty miles out of New Bedford and same distance from Nantucket. The season extends from June to September. The fish generally weigh 400-500 pounds and are from ten to twelve feet long. They are sold in New York.*" The Portland Historical Society in its publication *Proceedings of the Portland Society of Natural History* stated that this fleet operated mainly out of New Bedford; however, it is this author's opinion, as derived from various other sources, that the fleet itself was in large part native to Martha's Vineyard, the boats selling their stock to wholesalers in New Bedford. In those days, the bulk of the early catch was sold retail as fresh fish in New Bedford, Fall River, and Providence, with the remainder going to New York and to some extent to Boston. Later in the season, as the fish became more plentiful, the surplus of the catch was salted and exported to the West Indies and our southern states. The market in those years averaged four cents a pound for fresh fish; when salted the price was $6 per barrel or a drop of one cent per pound. It is fairly certain that whatever fish were directly landed at Martha's Vineyard were salted, as a fresh fish market of any importance could not have existed at that time on the island. At least two reports state that the fish when salted was considered to be very much a delicacy and far superior to other fish handled in that manner.

In rapid succession by the late 1860s a number of other ports began appearing in the fishery's history; Block Island seems to have been a seasonal center for boats operating from New London which fished in company with boats native to the island. Their catch was transshipped to both Newport and New London. Portland, Maine, also became an important fishing center. It was to Portland in 1865 that the vessel *Willy Smith* brought in a trip of forty-two fish which sold for $65 -- not much of an incentive for a continuing fishery, but that appears to have been a low time in the price structure.

An Established Fishery

By the 1870s, the industry was fairly well established although except in a few instances, it was not considered a primary fishery. It took some form of competing incentive for a medium-sized schooner to abandon a traditional fishery and enter the quest for swordfish. The most common incentive for changeover within the line trawl groundfish fleet was an abundance of dogfish sharks which during some years come onto the banks at approximately the same time that the swordfish arrive. The difficulty of working line trawls in areas where dogfish were abundant prodded many vessels to rig for swordfish harpooning, provided, of course, that the price justified the changeover.[1] Other vessels such as the larger line trawlers, which could by the fact of their longer range capabilities, evade the dogfish, would often carry harpoons and rig lookout towers in order to take advantage of any extraordinary concentration of swords which they might come upon. Thus the fishery of that era was well pursued even though on a short season basis and with many of the participants engaging in it only as the opportunity arose for sure success.

Harpoon Fishing Methods

By the latter half of the nineteenth century, the harpooning techniques of fishing for swordfish from Maine to Rhode Island had become relatively standardized. Most of the vessels which were engaged in this seasonal fishery were of the sloop and medium-sized bank schooner variety of up to fifty tons burden whose ordinary employments were either as mackerelmen, trawlers, or bank hookers. (Although some small sailing craft such as the catboats of Martha's Vineyard, Cape Cod, and Block Island also stalked swordfish, their landings were insignificant when compared against the whole.) Normally, most schooners and sloops of the size used as swordfishermen carried from one to five dories. Their crews varied from four to ten men. When engaged in their regular fishery, some were *"salters,"* but when swordfishing, most, especially following 1890, carried ice and landed their fish in a fresh state.

Essentially an inhabitant of the warm outer continental slope, swordfish come up on the cold northern banks only in a quest for the plentiful food resource which is present there. Most of this feed is layered near the bottom or in the mid-water range where the temperatures are the coldest. The heavy feeding and exposure to the cold waters produce a type of digestive state of lethargy wherein the numbed and belly-full swordfish surfaces to bask in the warmer upper surface layer while its stomach contents are being digested. In this state, it can easily be approached by surface craft.

As soon as the swordfisherman arrived on the banks, hands would go aloft with orders to keep their eyes peeled for the telltale dorsal and upper tail fin of a basking swordfish. When the lookouts sighted one of these finning fish, the harpooner would immediately take up position on a specially rigged bow stand located at the outer end of the vessel's bowsprit. When almost over the fish, the harpooner would thrust his shaft downward, driving in a detachable dart, to which was connected 400-900 feet of one-quarter-inch line. With the dart imbedded, the fish's lethargy would disappear, and in a burst of pain-driven power, it would streak away either horizontally or downward toward the bottom. As the harpoon line paid out to its bitter end, a keg float or skin-air bladder would be attached and thrown overboard. The purpose of this float was to absorb the furious surges of the fish. After the fish had tired, the float was retrieved, either by the vessel or by one of its dories launched especially for that purpose. Dories were most often used as they had the advantage of offering a less rigid platform from which to fight a fish; this meant less chance of the dart pulling out of the fish's back. Also, when the parent vessel's dories were utilized, more than one fish could be fought at a time, and this proved especially desirable whenever a large number of finning swordfish were in evidence. Whenever desirable weather and proper sea temperatures were present, the fish would be generally concentrated; at those times, it was advisable for a fishing skipper to harpoon as many fish as possible in as short a period of time. Without dories to fight each swordfish separately, the vessel would be restricted to limited harpooning opportunities. When first brought aboard, the fish were usually de-billed; but final dressing out and stowage awaited the end of each day's fishing which occurred when the fish began to sound, an event usually coinciding with the sun's lessening altitude by mid-afternoon.

The methodology described above is very similar to that which is still practiced by harpooners. The only major addition for the modern harpoon-equipped swordfisherman has been the use of radio-equipped airplanes to act as far-ranging fish spotters.

Early Grounds

In a Commissioner's Report of the United States Bureau of Fisheries, 1883, it was stated that *"Portland, Maine, is the headquarters for the fishery in eastern New England."* Mr. Whitten, then secretary of the Portland Fish Exchange, was quoted as estimating the number of fish landed there that year at two thousand, averaging each at 300 pounds dressed weight. In 1884, statistics tell us there was a total swordfish catch of seven thousand fish weighing in total approximately 2 million pounds and valued at $60,000 or three cents per pound. Portland led the landings with 1,062,500 pounds; Boston with 350,000 pounds; Newburyport and Portsmouth had landings of approximately 41,000 pounds; Gloucester and Rockport had 352,000 pounds; with New Bedford weighing in with 41,000 pounds. The fishery of that era was conducted more inshore than they would later become. In the last part of the nineteenth century, as well as in the early part of the twentieth century, much of the fishery was pursued inshore of the outer shelf as far north and east as the Bay of Fundy. Starting around the 1930s, except for warm water areas near No Mans Land off Martha's Vineyard and those waters adjacent to Montauk, Long Island, the harpoon fishery became mainly an outer bank enterprise. There is very little concrete information as to whether or not the fish had changed their habits of range or whether the more widespread inshore distribution was merely a result of a reduced resource. Individual swordfish seem to have an aversion to being in too close proximity to each other; thus when in large numbers, they may tend to spread out in their range, provided, of course, that water temperature and food conditions are acceptable. It is a rather big question as to whether climatic differences in water temperature could have been the sole cause of the early and wider distribution patterns. However, upon examining water temperature records for the inner Gulf of Maine (when the records became more plentiful as they did from the 1890s on) it is difficult to find any trends pointing to a radical water temperature difference from what we have today. Thus the wide distribution found in the 1880s would most likely have been the direct result of a much larger population of fish than was the case a few years later.

One of the best early grounds was inside Jeffrey's Ledge within the Gulf of Maine. Another, near Boone Island, was typical of the close inshore grounds which stretched as far east as Mount Desert, Maine. Fish were recorded as being taken well into the Bay of Fundy where a swordfish would be a rarity today. This is not to say that there were not fluctuations in the inshore availability of the fish during that period. In early September of

1883, it was reported that nearly a *"hundred vessels were cruising between Cape Ann and Grand Manan Bank."* Earlier that summer, the best grounds had been around inner Jeffrey's Ledge and off Wood Island, Maine, the latter within three to four miles of the mainland, but by August the fish had moved well off. In 1884, most of the catch was again taken offshore in the area between Jeffrey's Ledge and Mount Desert Rock, Maine. That season, according to A. Howard Clarke of the Bureau of Fisheries, there were few boats working on Georges Bank, *"where apparently a year or two earlier the fish had been plentiful."* In 1884, *"none but* Emma Clifton *fished there."*

By 1940, the literature tells us that it was a rarity to see a swordfish on the old inshore grounds. The shortage in 1940 was then attributed to the lack of mackerel on the grounds, but as we have seen since, years of relatively high mackerel abundance along the coast of Maine have not produced swordfish, except for an occasional stray. The same fluctuations in swordfish appearance can be applied to the more southerly waters close in to Block Island. In years past, swordfish were regularly harpooned within rowing distance of the harbor there; now to see fish that close to the island would be a rarity. In proximity to the Block Island area, summer water temperatures close in are certainly acceptable to swordfish, so there is no oceanographic reason why they should still not be common close to the shore. But they are not. The answer, without much doubt, is the reduced resource making the odds for an inshore appearance much less than they formerly were.

By the late 1890s, the fishery had taken on a somewhat different characteristic. It was mainly being conducted on the offshore banks and to a large extent within the greater New York bight[2] whenever temperature and feed became acceptable there. The area close to the now-named *'dumping grounds,'* being accessible to the fishermen of Martha's Vineyard, Block Island, and Connecticut, was a favorite and would remain so into the 1970s for the short-trip harpoon fleet. For the larger boats operating out of Newport and New Bedford, the southeast edge of Georges Bank was considered the best ground, although fish were often taken close in to Nantucket and even well into the South Channel. The drop-off of Georges Bank around Oceanographer's Canyon, north to the *'northeast edge'* was good from early July and on. The so-called *'Gulley'* located between Georges and Browns banks was also a good ground, but usually not until after mid-July.

By 1891 swordfish had begun gathering an enthusiastic reception in the marketplace, and prices had become more acceptable to the fishermen. There were some boats actively engaged throughout the entire season, June through October, but most of the production in that year can be attributed to ground fishermen and mackerelmen who were glad to take a respite from their regular fishery for a month or so.

It is quite difficult to compare the production of the period, 1890-1910, to that of later years. The swordfishermen of those days were severely handicapped by the absence of motor power in their schooners. With too much of a breeze (over about eight knots), it was difficult for a vessel to approach a fish as the approach speed was often excessive causing the fish to prematurely sound. A too-calm day, although ideal under motorized conditions, would place the sail-driven swordfishermen in an immobile state. Attempts were made to circumvent these calms by approaching fish in dories, but for some still-unexplained reason, it is extremely difficult to come up on a swordfish with a boat as small as a dory. Various reasons have been put forward as to why this is so, but the most generally accepted theory is that to a swordfish, a small boat probably resembles its natural enemy, the shark. So, when we read of records such as that made by the Gloucester schooner *Midnight,* of fourteen fish taken in one day on Georges in 1877, we can mentally allow a handicap; had she been working under power, her catch might have been doubled. The only worthwhile method we have by which to compare the fisheries of the 1890s and the early 1900s as against those of today is the average size of the fish caught. This the standard tool that most fishery biologists use to gauge the effect of any fishery against a given population. As a population begins to have inroads made upon it, the size of the harvested fish will decline. From 1880 to 1929, whenever records of weights for representative fish samples are of record, all indications are that the average size dropped from around 280 pounds to below 150 pounds. Then, from 1929 until the advent of the use of longline gear, weights held fairly steady. This would seem to indicate that although the cream had been skimmed; the fishery could be thought to be a resource sustaining itself. One cautionary note, though, regarding the weight reduction with the advent of longlining is that while the harpoon fishery was directed against large females on the shelf, longlining, being generally conducted off the banks over deeper water took in large part the smaller male fish. Nevertheless, the days of monster fish being landed with any frequency were over. It is interesting to look over the material which the *Proceedings of the Portland Society of Natural History* utilized for the study it published in 1947; within it is a list of large fish landed over the early years.

> A 16-foot swordfish landed in New York in 1791; one estimated to have been close to 800 pounds live weight which was taken back of the shoals right outside of Edgartown, Martha's Vineyard, in the late 1830s; an 850-pound fish, dressed weight, taken in 1894; two fish seen at Louisburg which dressed respectively at 915 and 903 pounds -- in the stomach of the smaller was found another swordfish *"about 3 feet long and apparently of about 18 pounds weight";* in 1917, a fish was taken off Cape Breton Island dressing 800 pounds; in 1921, Captain Irving King of the schooner *Higco* took a fish dressing at 915 pounds which was estimated at the time because of its condition, etc., to have weighed 1,160 pounds when it was harpooned off Georges Bank.

As the years went by, there were still a number of very large fish taken, but certainly not of the huge size or at the proportionate frequency with which they were recorded in earlier years.

In the early harpoon fishery, contrary to the situation beginning in the 1960s with the longline method and later in the 1980s with gill nets, it was a rare thing to see a small fish. A 30- or 40-pounder was an object of curiosity, usually well reported. One early evidence of a very small swordfish came to light in 1878: at the mouth of Gloucester Harbor a small *"mackerel shark"* (probably a mako or a porbeagle) was landed. In its nostril was found a swordfish bill detached from its owner; the young swordfish, as estimated by biologists of the day, *"could not have exceeded ten to twelve inches in length."* Rarely in the harpoon fishery were small fish taken or, for that matter, observed upon the banks. In the old days, fish landed that were less than 100 pounds usually came from line trawlers either fishing for halibut in the Gulley or off the edge of the banks. Quite a few came from tile fishermen fishing the 200-meter drop-off south of Nantucket. Since, until the beginning of the longline fishery in the 1960s, it was rare to land fish of less than fifty pounds, the fishery during the early part of this century was obviously harvesting a well grown out female class.

The early part of the twentieth century was generally a time of relatively high returns for those vessels fishing swordfish from late June through September. It was anticipated by the captains and their crews as a respite from their normal activity -- even being likened to a vacation when compared to ground fish line trawling and mackereling. The harpoon fishery was (and still is) the closest thing that a commercial fisherman ever sees to sport. To a doryman fighting a large fish, it could approximate the experiences of a big-game hunter facing a charging rhino.

Danger from fish charging a dory or the weather of the offshore grounds were not the only dangers swordfisherman faced. In the summer of 1918, a German submarine surfaced among the United States and Canadian fishing fleets working the offshore banks. The skippers were given a warning to launch their dories whereupon the fishing vessels were sunk by gunfire or demolition -- eighteen were sent to the bottom. All were schooners, most being swordfishermen. Twenty-four years later, in another war, a motorized swordfish schooner was sunk by U-boat shellfire.[3]

The Fishery Changes

Following 1929, which has come to be known as a bonanza year in the harpoon swordfishery, a great deal of interest developed over extending the normal season through exploration. The forerunner of such exploration attempts was carried on by the schooner *Liberty* of Boston. In April, she sailed for Martha's Vineyard, from there heading south to fish in the vicinity of Cape Hatteras, *"a rather unusual proceeding."* *Liberty,* under Captain Claude Wagner, apparently found no fish surfacing off Hatteras.[4] It is not known how long Wagner remained in the Cape Hatteras area. There is a report that on 31 May 1930 he landed a fare of three fish taken *"sixty miles south southeast of No Mans Land"* which he brought into New Bedford where they sold for the then almost record-breaking price of seventy cents per pound. From this information, it seems likely that Wagner probably surveyed southern waters for no more than a month (if one calculates travel time by sail). This was not the first exploration southward as the Boston Fish Bureau in January of 1922 had receipts of fresh swordfish shipped from New Jersey. The Bureau's report stated at the time, *"something unusual in the receipts at this season of the year has been the arrival recently of several swordfish from the south, reported to have sold at 45 to 50 cents per pound."* By 1930, a great deal was becoming known as to when the fish might be expected to arrive on any given ground. It was generally theorized that from the last week in May to June first, fish coming in from deep water to the south would arrive off Block Island and No Mans Land, from there to *"gradually approach the inside waters of Vineyard Sound."* By the second week in July they could be expected to leave, traveling in a southerly direction to the outside edge of Little Georges Bank and the southeast edge of Georges Bank. The fish would usually remain awhile at the *"southeast corner of Georges from whence they rapidly disappeared."*[5] This was the group of fish targeted by all but the largest vessels, and it was never thought to have been as large as the populations of fish that came in farther to the eastward at the Gulley, and on the banks off the Canadian Maritimes.

Traditionally, the appearance of the main population of swordfish could be expected on Georges Bank by the first week following the Fourth of July. The best ground was then thought to be in the vicinity of 65°W and 41°N; there the fish seemed to arrive earliest, and it was always considered to be a good area for the remainder of the season. The modern harpooners of the 1970s, as witnessed by this author, seemed to agree, as a great part of their

activity encompassed the area from those coordinates and beyond in a northeasterly direction and extending somewhat to the north of Corsair Canyon.

The reasons may be varied as to why Georges Bank attracts swordfish. Generally, it is because certain parts of the bank's outer edge gets more than the average share of warm water influence from the Gulf Stream. There are also some strong upwellings along the edge of Georges Bank, particularly in the area of Corsair Canyon. The general rule for those looking for surfaced swordfish was to stay just inside of the bank's edge (that is slightly inshore of 200 meters). However, fares were often taken by harpooners looking for blue water when the Gulf Stream's boundary was well offshore of the bank and in water considerably deeper than 200 meters. It is reported that in 1916 the schooners *Barbara* and the *Bernie and Bessie* took between them eighty-seven fish on 15 July 1916, at a position well offshore of the 200-meter curve. There is no doubt that the fishermen of that time realized a relationship between the edge of warm oceanic water and the cold water on the banks and that they would try as best they could to locate these edges when looking for swordfish. It was generally accepted that when the warm water was too far off the bank, prospects generally were not good. When at any significant distance off the bank, fish rarely surface, and when they do, they usually are wild and hard to approach; 15 July 1916 was apparently a variance from the general rule.

In the 1930s, it was generally considered that 75 percent of the catch would be taken on Georges Bank or over the deeper water which divides Georges Bank from Browns Bank off Nova Scotia, making up the difference. The deeper water which separate Georges Bank and Browns Bank is named the Northeast Channel on charts, but to the fishermen, it is known as *"the Gulley."*

Canadian Ascendancy

The Nova Scotian banks were initially, if only casually, fished for swordfish by the American fleet as early as the 1880s. The Canadians themselves did not enter the swordfishery in any significant way until 1909. In that year they began reporting swordfish landings as a separate entry within their fishery statistics, although it was then of little importance when measured against their overall fishery.

Prior to the 1930s the American vessels fishing the Canadian banks brought their fish into American ports. After that time most of their landings were made in Nova Scotia with the fish being transshipped by train, or in some cases carrier vessels, to the port of Boston.[6] The Nova Scotian offloading ports generally used were either Louisburg or Yarmouth. The Canadian fishery was more incidental than was its American counterpart. As late as 1929 Canadian landings were still less than one million pounds compared to over five and a half times that amount by the American fleet. By 1935, the Canadian catch had risen slightly while the American catch had dropped over the four preceding years.. (See graph entitled *Swordfish Landings, 1909-1959, United States and Canada.*) By 1951 (which marked a reduced point in United States landings to only 162,000 pounds) the Canadians produced over 2.5 million pounds. From then on, the Canadian catch soared upwards; in 1959, their production was 6.7 million pounds while the United States production was only slightly higher than half a million pounds.

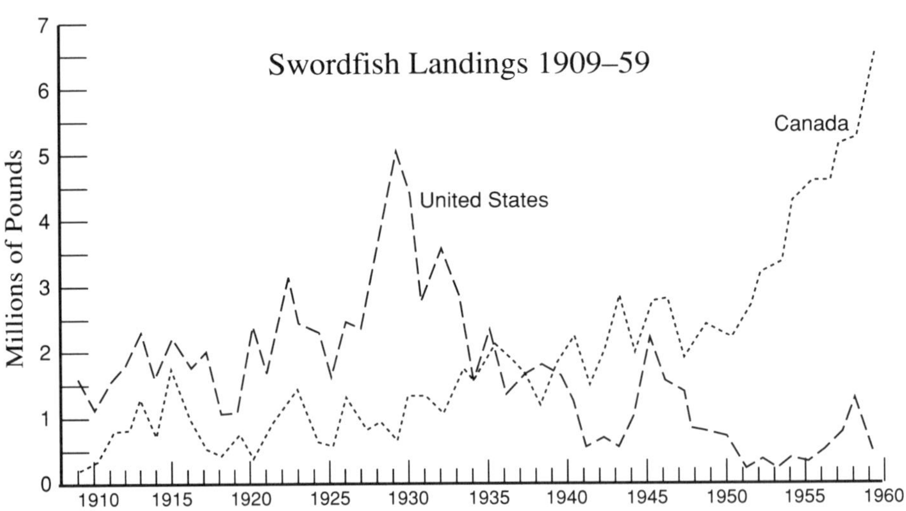

Swordfish Landings in Canada and the United States.

The decline in United States swordfish production which began from 1931 to the early 1960s, at least according to most authorities, was not attributable to a lesser supply of fish. The overall population, as judged from Canadian records on all grounds, remained good. Observations by American skippers, as well as Canadian skippers fishing Georges Bank, relate that the supply was relatively constant there throughout. The reason for the decline in the United States catch has assumed to have been the negative difference in profitability for the fishery when weighed against other fisheries. According to Tibbo, Day, and Doucet, writing in 1961, the Canadians' simultaneous gain in swordfish production was principally attributable to the development of a small tonnage line trawl fleet designed primarily for the groundfishery but which was easily converted for harpooning. These vessels, generally in the 40- to 60-foot class, were seaworthy, fairly long ranged, and economical to operate. Swordfish harpooning, with the increasing price for the product, was seasonally attractive to the Canadians who historically experienced annual price drops for groundfish landings during the months of July and August. These were the same months when swordfishing was at its best and when the price was dependably consistent.

As to United States production during the period 1909-1959, one can make some relevant observations: The swordfishery was relatively steady and profitable for the United States fleet until about 1927. Although peaks and valleys of production can be found in any fishery, in the case of the swordfishery, such peaks and valleys were most likely attributable to oceanographic conditions and technological change. In 1927, the fishery began to have noticeably increased landings. The best explanation for that was the industry's trend toward conversion to auxiliary power plants for vessels that formerly were propelled by sail alone. This made the swordfishery definitely more efficient. The crews of auxiliary-powered sail vessels could now leave port with better prospects of more days to be spent actually fishing. The modernization was well along and in time to take advantage of favorable oceanographic conditions which produced the high landings experienced in 1929. As earlier stated, beginning with 1931, landings began to slowly decrease, a direct result of the changeover from the New England bank schooner to the modern otter trawler. The owner of an otter trawler probably found it more profitable to fish year round for the groundfish species. The decline in landings which began in 1931 continued, with a drastic dip brought on by the war dangers of offshore fishing during 1942 and 1943.[7]

Landings rose again following the conclusion of the war. The postwar rise in swordfish production was a phenomenon occurring as well in a number of other fisheries, a result of more plentiful labor force as well as the addition of new vessels, some of which were military surplus. The partial resting of fishery resources during the war years may also have been a factor. In the swordfishery, the ratio of price, when compared against that for groundfish, was beginning to become attractive. For those summer, part-time fishermen, a swordfish boat was cheaper to equip than was an otter trawler. Also, the hunting appeal of the swordfishery was a strong influence to many. Following the 1947 season, landings began to decline, this time, it is believed, because of the dropout by many of the former servicemen who had realized that the swordfishery could be rife with uncertainty in production as well as being inefficient of labor when compared to otter trawling. Swordfish landings reached a low point in 1951.

Because of a reduced catch effort which had begun in the late 1940s, the 1960 swordfish resource was probably in a much more improved state than it had been during the 1930s. The record clearly shows that it had only been moderately and spasmodically fished since the heavy efforts of the late 1920s and early 1930s. It is true there had been some entry onto Georges Bank by Canadian vessels, but this had not been overly significant as the majority of Canadian effort had been expended on the banks offshore of Newfoundland.[8] By 1960, the price was high enough to make the fishing again appear profitable for the United States fleet. A number of American fishermen from Portland to New Bedford were thinking of entering. The fly in the ointment was that the Japanese had begun exporting swordfish produced from their Pacific longline fisheries, and these were just starting to impact the United States market. This would become an increasing worry, not only to the Americans but also to the Canadians whose major market was the United States. There was, though, the belief among some fishermen and dealers, as well as marketing specialists from government, that the domestic market had hardly been tapped. Swordfish as a food item was still largely unknown and was sold only within proximity of the major landing ports. The fishery's major problem was that the catching methods then being employed by the Americans and the Canadians (harpooning) was extremely weather dependent as well as labor intensive. As the seasonal fishery was then being carried out, production depended upon relatively smooth seas, correct water temperature conditions up on the banks, and reasonably good visibility. These conditions, when taken together, are not very common north of Cape Hatteras even during the summer. In order to circumvent these limitations, some thought was being directed toward the Japanese technique of longlining, then a major catch method for tuna and for billfish.

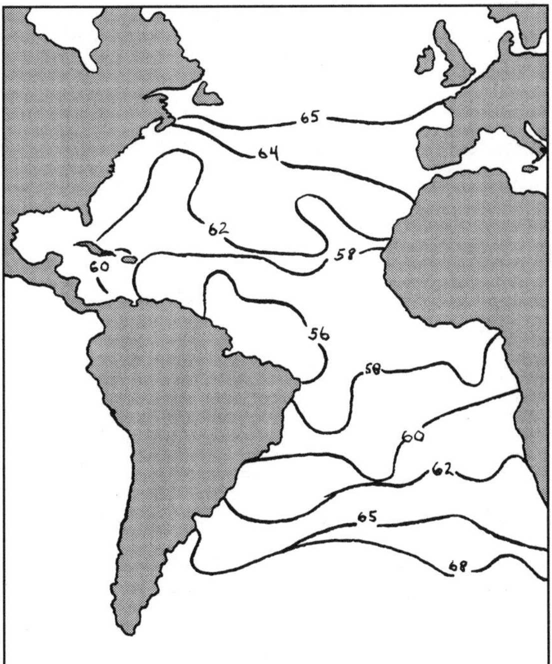

The Japanese longline fishing grounds in the Atlantic Ocean as they expanded on a yearly basis from 1956 through 1968. This graph represents both the billfish and tuna catches; however, it is useful in illustrating the extent of the areas sampled by the Japanese.

Beginnings of Longlining in the 1960s

The longline was undisputedly the only feasible method being used during the 1960s for the large-scale commercial harvest of billfish. There was nothing new in the employment of free floating lines to catch pelagic fish. The Cubans, fishing near their island from small vessels had been targeting billfish for years, their gear being restricted in size, though, to 4-hook sections laid across the axis of the current. The Japanese originated the oceanic longline concept employing gear consisting of up to 6,000 hooks suspended from a continuous main line, some main lines being forty miles in length. The Japanese longline fisheries for billfish and tuna had proved to be highly successful, and by 1959, Japanese vessels were operating in the eastern Pacific, coming within soundings of the Mexican coast. Within a matter of a few months, the Japanese had extended their efforts into the Atlantic, soon to be joined by the Taiwanese and the Koreans.

In 1959, Canada and United States fishermen were aware of the Japanese longline fishery, but significant interest to compete with them was not being shown. In the United States, the first serious consideration to pelagic longlining was an investigative longline trip made in June 1959 by a trawler out of Gloucester which streamed a surface-floated line to the seaward of Georges Bank. This vessel, *Golden Eagle,* owned by Arnold Vogle of Riviera Packing Company, Eastport, Maine, landed what was the first United States's commercially longlined catch of tuna taken in the Atlantic. The vessel's skipper was Captain Edward Fragata. The newspaper accounts in disclosing particulars described the trip's sponsors as Vogle, who had provided the vessel, and the Bureau of Commercial Fisheries, which had supplied the fishing gear. The *Golden Eagle's* trip itself consisted of nine days of actual fishing during which a total of twelve miles of gear was streamed during daylight hours. The catch consisted of 320 blue fins and 10 yellow fins for a total of 35,000 pounds. Although the catch was good for an initial trial, Vogle felt it was not promising enough to justify removing *Golden Eagle* from the ground fishery in which she had been doing reasonably well. Additionally, the marketing of tuna had its problems which were a bit too difficult to solve at the time. Interest in longlining waned, but only temporarily.

In May 1961, with Frank J. Mather III aboard as senior scientist, the Woods Hole Oceanographic Institution's vessel *R/V Crawford* left its dock outbound out for Hudson Canyon (south of New York) to prospect for tuna. The catch made on that first trip was not spectacular, but the experience provided Mather with a lot of information, the most significant item being that the best catches seemed to occur when the sets were made on or around the 500-meter curve. Several trips followed, all adding to the knowledge of gear and techniques, but nothing occurred to indicate the possibility of a latent swordfish resource. The gear was soaked only during daylight hours which is a time of day unsuitable to feeding swordfish. Later, while on another longline exploration off the Virginia Capes, R/V *Crawford,* was forced, due to heavy weather, to leave her gear. Upon returning to it three days later, she hauled back finding the shark-bitten remains of sixteen swordfish, thus giving a clear indication that swordfish were in those waters and were thought to be night feeders. Thenceforth there was some emphasis placed by Mather on swordfish, although tuna continued to be *Crawford's* major quest.

The next major point of interest took place during the summer of 1961 when Norwegians, operating off the Canadian and United States east coasts, began conducting a productive fishery for porbeagle shark *(Lamna nasus).* They had promoted and developed a market in Italy where this variety of shark was sold as a gourmet item.[9] The Norwegian enterprise was composed of a number of longliners, which operated in the Gulf of Maine and along the shelf edge between Georges Bank and the Grand Banks. They were temporarily homeported at the French islands of St. Pierre and Miquelon, where they transferred their catches to one of two motherships for transshipment to Italy. On 31 July, one of the Norwegian vessels, *Tampen,* made port for repairs at Gloucester, Massachusetts. The word got around that she had on board four tons of swordfish which were taken incidental to her catch for porbeagles. This was probably the first direct knowledge that American fishermen had of the Norwegian longline operations being conducted off the United States coast. Such an ignorance was not the case with the Canadians who had spent that summer watching swordfish coming over the Norwegians' railings. What the Canadians had seen at sea, combined with what the Americans had seen in Gloucester, was fuel for a lot of talk over that winter. Much of that talk took place around the fish houses and bars of the fishing ports, but some of it occurred in the offices of fishery bureaucrats, both in Canada and the United States. The Canadians were apparently the first to react, as their Department of Trade and Industry chartered a groundfish line trawler named *George and Pauline* which they scheduled to start operations by July of the coming summer.

George and Pauline, on its first trip of that summer (1962), took sixteen swordfish. Using Norwegian-type shark gear, 6-foot gangions were hung from the mainline which in turn was suspended but a foot or two beneath the surface. The researchers were enthusiastic and so were those fishermen who witnessed the off-loadings. By early September, the Canadian fishing industry had ten large halibut-type trawlers at sea, all modified for swordfishing. An unknown number of smaller tub trawl vessels were also longlining. The best reported trip for one individual Canadian vessel that summer was 100 fish taken in four days; the largest trip was 184 fish, but that was from a considerably longer time spent at sea. The longline, as it came into use by Americans and Canadians in the early 1960s, consisted of a sub-surface free floating nylon line of usually ¼" diameter. To this was attached at 60-foot intervals gangions of 10 to 30 feet, baited with fish or squid. The longline itself was buoyed at intervals with Radar reflectors at one mile intervals. American boats fished from 5 to 25 miles in one longline from which were attached up to 2500 hooks. At the onset, of the fishery, few boats set more than five miles of longline. As experience was gained, longer gear was employed.

From Portland, Maine, steamed the 102' side dragger *Gulf Stream,* skippered by Captain Richard Paulson. Paulson went east to the Grand Banks where, stalking swordfish, he took 35 fish. Most were taken by harpoon, but one longline set was successful. Paulson must have learned something from that set, as when his next trip ended on 18 September, he landed 119 large swordfish. All had been taken by longline.

At Woods Hole, Massachusetts, Henry W. Klimm, one of the more innovative captains on the New England coast, and the owner of *Cap'n Bill III*, was busy cooking up a cooperative deal with the United States Bureau of Commercial Fisheries and Woods Hole Oceanographic Institution (WHOI). Both the WHOI and the Bureau people wanted production data on the then unknown tuna resource. Klimm did, too; but he also was curious about the swordfish potential. The deal worked out between Klimm and the Bureau was that Klimm was to supply his vessel, its crew, and fuel. The Bureau would supply the longline gear. The vessel would keep all the fish that were caught. WHOI would put aboard a research assistant by the name of Martin Bartlett whose function would be to advise and record. Before *Cap'n Bill III* left the dock on 8 October, Klimm and Bartlett agreed that at least one or two night sets should be made for swordfish; the night sets were to be the initial ones.

Klimm returned to Woods Hole after seven days and five sets with 34 swordfish and 69 tuna. The gear they used was crude, and nobody aboard knew the right way to fish it; but as one crewman put it, *"by God, they had caught swordfish."* Everyone aboard had the fever! Adding to their exhilaration was the news that Paulson had again brought his *Gulf Stream* into Portland, this time with 105 fish weighing 21,000 pounds. He had taken them off the Canadian St. Pierre Banks. Klimm and Bartlett, now completely enamored with the prospects of a huge swordfish resource, speculated that the swordfish's distribution range must be the entire outer shelf, a distance of over at least the 1,000 miles as measured from Georges Bank to the Grand Banks. Klimm and his crew, together with Bartlett, worked almost around the clock refining and adding to their gear. By 23 October, they were ready to go out again.

On 31 October, *Cap'n Bill III* was back in port with 101 swordfish and 145 tuna (mostly blue fins).

On its third trip, *Cap'n Bill III* encountered bad weather. Two gales came in succession, one with such force that while lying hove-to, they were blown seventy miles to the southward over a 24-hour period. Despite the weather, Klimm fished whenever he could; when he did, it was productive. The score was ninety-four swordfish.

Around this time, a third American boat, fishing with only two to three miles of longline, and hauling that back by hand, landed sixteen swordfish at Point Pleasant, New Jersey. The boat was *Elsa C* with Nils Rasmussen aboard as skipper.

Paulson's *Gulf Stream* was out again. Because of the oncoming fall weather which is notoriously bad on the Canadian banks, Paulson planned this time to work more to the westward off of Georges Bank. Setting over Hydrographer Canyon with 550 hooks, he took thirteen swordfish on his first try. Working farther to the westward he consistently caught fish on each night. His eighth and final set was made over Hudson Canyon; from there he proceeded into New York Harbor with eighty-eight swordfish and eighteen tuna. Finding the price unsatisfactory at Fulton Fish Market, he headed to the east up Long Island Sound to Newport where he sold his fish. It was of interest to Paulson and to Klimm as they talked by radio that the blue sharks seemed to be less plentiful than they had been. Apparently they had moved offshore; Paulson had caught only five of them on his last trip. All in all, everything was coming up roses -- a high fall-season swordfish population with few sharks to make life difficult.

On 2 December, *Cap'n Bill III* came in with a fourth trip, this time from Lydonia Canyon. The catch after five sets: eighty-four swordfish, one albacore *(Thunnus alalunga),* two big-eye tuna *(Thunnus obesus),* thirty-four blue fin tuna, two blue sharks, and two mako sharks. Water temperatures had averaged 58° to 60°.

Three days later, *Elsa C,* which had been fishing Hudson Canyon, docked at Point Pleasant with twenty-five swordfish and one mako shark. The swordfish averaged 180 pounds each. The crew, after hauling the longline by hand, was in a state of near exhaustion.

On 21 December, *Cap'n Bill III*, back from her fifth trip, landed 180 swordfish and 44 tuna at the Woods Hole dock. Klimm and his crew padlocked the hatch and walked away. Two days later, with their consignment of 180 fish successfully routed through the marketplace, they returned to *Cap'n Bill III* and swung another 186 fish onto the dock. Klimm, a fisherman with long experience, knew full well that if he had initially dumped his entire catch of 366 fish into the laps of dealers not previously used to handling fresh swordfish in high numbers, only a flooded market and low prices would have been the result. Due to Klimm's astuteness, the market held, and he and his crew grossed $31,440. This was probably a record for dollar value in New England history for any one trip of short duration up to that time. Productionwise, they had made ten sets which averaged out to 6.4 swordfish per one hundred hooks. To fishermen, the swordfish grounds were beginning to be like the Klondike had been to gold miners. In New Bedford, the crews of two vessels, the *Sunapee* and the *Josephine and Mary,* were busy fitting out for longlining while others were planning on it.

The rapid increase of the Canadian longline fleet was showing few signs of leveling off. Further growth was spurred on by the catch of *Dorothea Reeves;* in September, under the command of George Kendall, she off-loaded 101 large swordfish at Sydney. By the end of 1962, it was estimated that at least thirty-two Canadian longline vessels were in the swordfishery. They had caught upwards of 1,500 fish amounting, by estimate, to approximately 400,000 pounds. The gross price to the Canadian boats had been good; but because of the extra transportation costs incurred for shipping to the markets of Boston and New York, it did not come up to the seventy cents average per pound then being received by American boats. For gear, the Canadians were using halibut trawl lines which had been modified to float free at the surface. Herring was the most common bait being used.

Thanks to the logbooks supplied by Bartlett to *Cap'n Bill III*, *Elsa C,* and *Gulf Stream,* and the cooperation of their skippers in filling them out, we know a lot about the catching methodology used in those early months. Those records also make us aware of the composition of the catches as well as the average weights of the fish taken. *Gulf Stream,* for instance, had the most advanced gear including a homemade wooden spool arrangement for rolling up the mainline -- the line being first hauled in with a patent Japanese longline hauler, and then wound onto the spool. *Cap'n Bill III* was also equipped with the Japanese-type hauler, but after coming aboard, the line was coiled into tubs by hand. Paulson used roughly 500 hooks on each of his sets and attached his gangions to the mainline with snap-on swivels. For bait, he used herring or mackerel. Klimm on the *Cap'n Bill III* dragged for his bait using, in his words, *"hake, butterfish, eel pouts, and anything else which was about a foot long and shiny."* *Cap'n Bill III's* tub gear had about 200 meters of mainline per tub to which gangions were attached at 20-meter intervals. Unlike Paulson's snap-on devices, Klimm tied his gangions permanently onto the mainline. As he streamed the mainline, and each tub's contents went over the stern, the next tub was tied on. To each tub length, there was connected an inner tube float tethered to the mainline. Buoy lines for the floats were in lengths of from two to twenty meters, depending on what depth it was felt the fish might be feeding. Experimentation, however, did not seem to point to any noticeable depth preferences on the part of the fish. Both Klimm and Bartlett believed that the longline generally took fish equally well at all the depth settings which they tried.

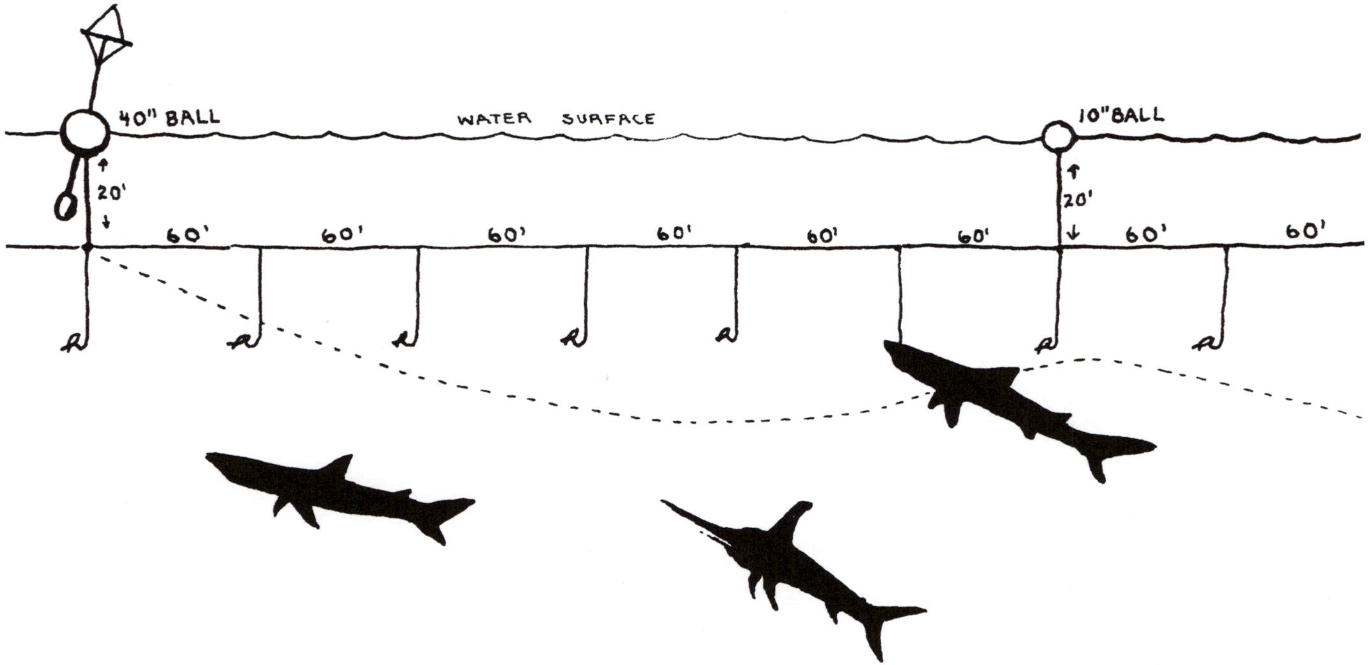

A typical longline set (circa 1960s and early 1970s). Sketch illustrates spacing of the gear -- the dotted line indicating the catenary which the sections take on shortly after the gear is set out. The catenary tends to increase the longer gear is in the water. It can increase dramatically from the weight of fish on any given section or from eddy currents drawing the floats together.

Reading over the logs of that first longline year, it is interesting to note how crude the search techniques were when compared against those later practiced.[10] As the year came to an end, Bartlett's analysis indicated that the fish were averaging about one hundred pounds each, noticeably less than those taken in the harpoon fishery. Although the significance of water temperature was known to both Klimm and to Paulson, neither of them seemed to place any great importance at that time on mid-water thermoclines or on vertical thermal fronts.

The Boom

For the swordfishery, probably the most noteworthy event of early 1963 was held at the Free Public Library in New Bedford when more than 200 people showed up for a program on the discussion of the swordfishery. It was sponsored by the New Bedford Seafood Producers Association. The organizers were specialists from the United States Bureau of Commercial Fisheries who were at somewhat of a nonplus over the audience's quest for investment advice. The Bureau's team leader, Peter Wilson, started off with the word of caution: *Cap'n Bill III* and *Gulf Stream* had landed over $100,000 in stock so far while another vessel, *Gertrude D,* had made a 9-day trip stocking absolutely nothing. Wilson explained how research done in the past had been insufficient to establish any resource evaluation concerning the size of the resource. However, things did look very promising, not only for the swordfishery but also for tuna, the latter discussed either as an incidental catch or as one to be fished for separately. Answering a direct question from the floor as to the size of the swordfish resource, Wilson answered, *"I repeat...there is no guarantee of a resource of commercial magnitude waiting to be tapped. We think there is, but we have no way of being sure about it."* Wilson then went on to discuss the swordfish's probable affinity to water temperatures and how this could be an important factor in locating fish. Estimates were given as to how much it would cost to outfit an existing fishing vessel for entry into the longline fishery, $5,000 being the round figure that was given. In answer to a direct question as to how to go about making longline sets, Wilson said that this was one of the things that the Bureau was trying to tie down. He mentioned *Rorqual,* a research vessel, which during the last summer had made six sets in about fifty meters of water south of No Mans Land. When *Rorqual* made her sets, there had been extensive finning of swordfish in the area. Yet only two fish were caught by longline. Wilson said that even if fish were known to be in an area this was no guarantee of catching them by hook. Wilson concluded by stating that many questions must be answered before any guarantees could be made.

The factor that really had everybody's interest was the possibility that the swordfish's range was much larger than previously believed. Stories about the fish were coming in from everywhere: In the 19 March edition of a South Carolina newspaper, there was a piece regarding the captain of the Charleston bar pilot boat, who, while standing off at the end of the harbor jetties, hooked onto and landed a seven-foot swordfish. Although a fishery research vessel had earlier taken two swordfish at the edge of the Gulf Stream directly off Charleston and a snapper fisherman had reported taking a fish along the 50-meter curve in 1962, no one in southern waters had ever heard of a swordfish being caught as close inshore as the one taken by the pilot boat. *Xiphias gladius* seemed to be everywhere.

Fishermen from Oregon to Newfoundland were putting to sea with tubs of gear and great piles of buoys and inner tubes jamming their work decks. By April, the Bureau of Commercial Fisheries reported seven vessels working the Norfolk Canyon area alone. The seven were out of New York and New England ports, but the report anticipated that an eighth vessel, a local Hampton, Virginia, boat named *Leslie H,* would soon join them.

In early 1963 (January), *Cap'n Bill III* had already made two trips to Veatches, Corsair, and Hudson canyons. Neither trip was highly successful as between both trips only forty-five swordfish were landed. Also back out was *Gulf Stream,* which boated twenty-eight on a trip during January. Nils Rasmussen's *Elsa C* made two trips that month catching no swordfish on the first one but seven on the second. Rasmussen went back out, this time to the Hudson Canyon area and came up empty. Temporarily discouraged, *Gulf Stream* and *Elsa C* laid at the dock. Henry

Klimm of *Cap'n Bill III,* after discussions with Frank Mather, Bartlett, and others, was determined that the fish were there waiting to be tracked down by following the correct temperature gradients. Swordfish had been found off Cape Hatteras as early as the winter of 1961 by *R/V Crawford,* so despite the terrible winter weather, Klimm kept trying. After one so-so trip off the Delaware Capes, he took his *Cap'n Bill III* east of Cape Hatteras and there he hit the mother lode, but with problems: lots of sharks and wildly erratic currents, the velocity and directions of which appeared impossible to predict. It is in the area off Cape Hatteras that the Gulf Stream shifts inshore and offshore most rapidly, and the shifts happen with no predictability. During the winter and spring short-lived eddies bulge out from the western wall only to retreat again into the main flow; these eddies acting like great swirling hydraulic vacuums which can suck away the longline gear of any fisherman bold enough to set. Under such circumstances, a longline fisherman has no real estimate of his gear's drift rate or its direction. For Klimm, this translated into a lot of lost longline both from the currents and from the sharks. Still he took fish. Bartlett and Lineweaver, writing in the *National Fisherman* the next spring claimed that by midwinter, *Cap'n Bill III's* gear had suffered losses to the tune of $3,500; but what wasn't lost had upon it a catch of up to sixty-five swordfish a trip. Despite the handicaps, the hook rate efficiency was still better than that of the past fall when the catch was being taken from the offshore canyons to the north.

Klimm developed a system which basically was to work a little deeper than the edge of the 200-meter curve, making his sets whenever thermal fronts slightly colder than 60° F could be found. As long as depth and temperature conditions remained as they had been at the time of setting, things usually went well. However, if the Gulf Stream moved in during the night, with it usually came the sharks, which -- together with the effects of the current -- parted, tangled, snarled, and generally turned the longline gear into one great frustrating mess. Hardly discouraged by such setbacks, Klimm and his crew kept at it and made money while improving their techniques. Klimm fished less gear than he had that past summer up north. He had to in order to minimize his gear losses.

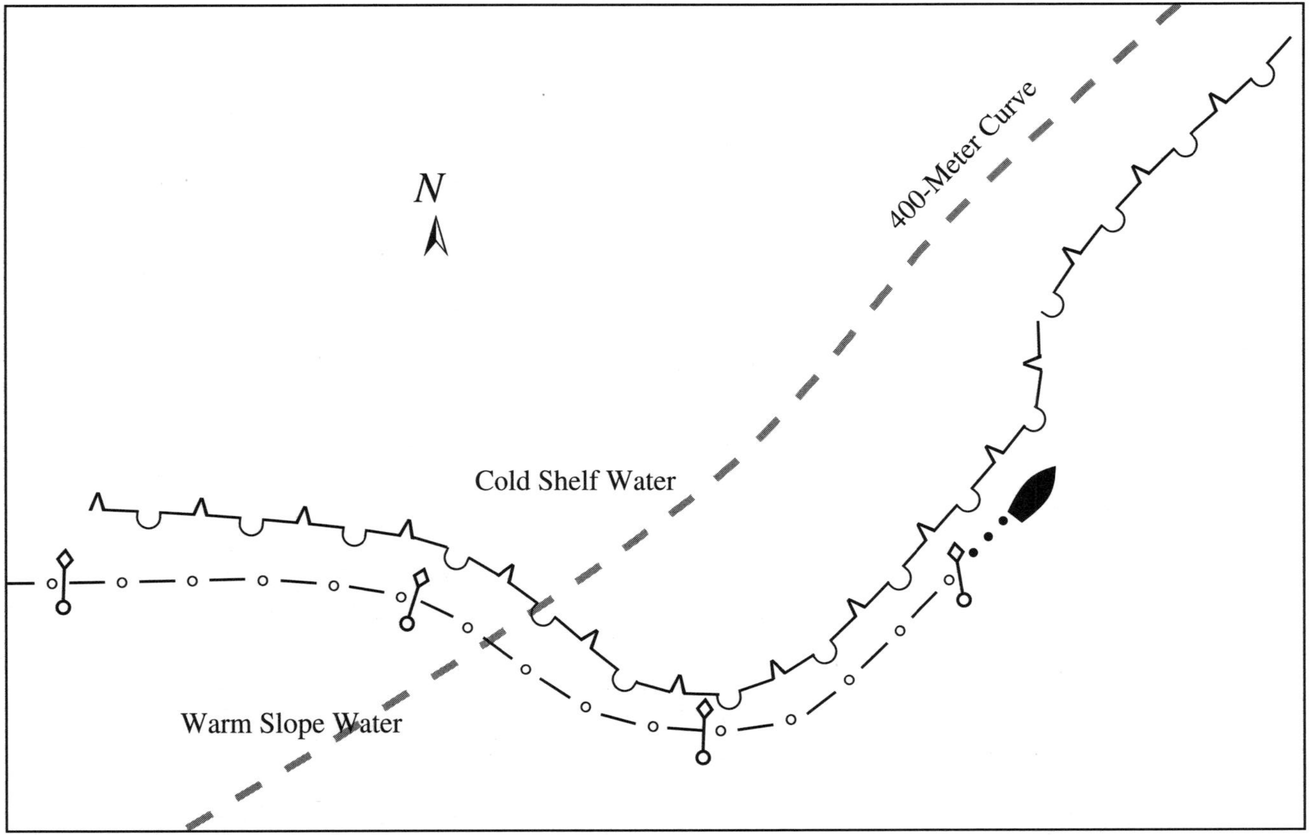

Typical method of setting out longline gear (free floating from vessel). Hypothetical location off New England states.

Indications to those active in the fishery were that given seaworthy vessels with competent and tough crews, the swordfishery could become a year-round business.

With the cessation of the winter storms and the coming of warming temperatures, things began improving oceanographically; yet new problems brought on by overcrowding on the grounds began to arise. It was not out of the ordinary, when setting out at night, for crews to see the lights of six, twelve, or even fifteen other boats, all of them streaming gear. Whenever a skipper set his gear too close to that of another, the result could be a drifting together of the mainlines during the night. The result was hours spent the next day untangling the resulting snarls and taking away from time which otherwise would have been spent getting needed sleep. Operationally difficult, but not to a critical point -- at least not at a time when most crews, even first trippers, were making money.

Notwithstanding the relatively high cost of equipage, entry into the fishery remained continuous, encouraged by the growing belief that this was to be a year-round fishery. For the larger vessel which could go as far north as the Grand Banks, this seemed a reasonable conclusion. Past experiences of the Canadian harpoon fleet as well as observations made by others pointed toward a large fish population on or near to the Grand Banks and which stayed there well into the fall. Besides that, Klimm had proven that a winter fishery existed to the south off Cape Hatteras.

The conversions of vessels from other forms of fishing resulted in busy times for the shipyards and the fabricating shops of Gloucester, New Bedford, Hampton, and Cape May. Line haulers, tub racks, and experimental type longline reels needed to be fabricated and installed. Cordage manufacturers were back-ordered for mainline and gangion materials.

Meanwhile, in the marketplace more than 100,000 pounds of swordfish came into Fulton Fish Market alone during the first two weeks in July. The last half of that month started off with 41,400 pounds being received during just one day. At Boston, 500,000 pounds was reported for the week ending 12 July. The increase in production, as compared to 1962, was dramatic. The Fulton Fish Market receipts in June 1962 had been 30,900 pounds, while in 1963 for the same month, they came to 94,700 pounds. Putting things into comparison for the first six months of 1963, Fulton receipts were 173,200 pounds as compared to 34,300 pounds for the same period in 1962. Prices extending into the middle of June of 1963 remained at a fairly stabile level of between sixty-five cents and eighty-five cents a pound. However, by the last half of June, things began to change as the rule of supply and demand took over. Prices at Fulton fell to thirty-five cents and then to twenty-eight cents.

Among the press reports dealing with this newest fishery went a lot of misinformation. An example of this was a story in a Nova Scotia newspaper, datelined 16 July 1963, giving the report of a 280-pound female swordfish caught off Shelburne, Nova Scotia, which had fully developed spawn. According to the reporter, *"this uncovered a secret that has puzzled fishery scientists for years."* The writer, who no doubt was getting his information from a fisherman rather than a scientist, believed this solved one of the mysteries of the sea, because, *"for years, scientists have been searching for the way swordfish multiply for it was not known if they spawned or pupped as do shark, porpoise, and dogfish."* The *"discovery,"* according to the reporter, elated fishermen *"over the knowledge which had been kept a secret in the ocean depths for centuries, as it means that the swordfish is similar to the cod and haddock in multiplying. That means that fishing for the broadbill by trawl or longline will not diminish the resource."* The truth of the matter was that scientists for decades had known that swordfish were egg layers, a knowledge also shared by the better-informed fishermen.

For the time being, some fishermen seemed on their way to riches. Boat owners, busy with the process of rigging up, were spending money, and the supportive industries were clearly benefiting. An example of this correlating benefit was described by Arthur O'Leary, president of Lambeth Corporation of New Bedford, who gave an interview for an article in the *New Bedford Standard-Times.* Mr. O'Leary, caught up in the boom mood, painted a glowing, albeit still to be proven picture when the article quoted him as saying, *"further research indicates that the entire*

Atlantic is loaded with broadbills, and that they can be obtained all year long."

Even some of the sportfishermen were uncharacteristically optimistic over the *"inexhaustible"* swordfish supply. John Vallone no doubt was one of them. When fishing on the charter boat *Mako II* out of Rhode Island, he boated a 297-pound broadbill winning the Cuttyhunk Derby. A few of the sportsmen, more in character and predicting doomsday, were beginning to worry whether, in light of the longliners, any swordfish would be left for them.

By August, the edge of the outer shelf was crowded with American and Canadian longliners. Something had to give. The resource, whenever space could be found to fish for it, seemed to be still abundant, too abundant in fact for the good of the marketplace. Prices kept getting worse. Moses Schonfeld, owner of *Debbie and Jo-Ann,* a longliner out of New Bedford, buttonholed United States Senator Leverett Saltonstall. Saltonstall, in a follow-up press release, claimed that *"the American swordfish business is infused with disaster due to Canadian subsidized vessels dumping large quantities [of swordfish] in American ports."*[11] The Senator, further inquiring for facts from the Bureau of Commercial Fisheries (soon to be renamed the National Marine Fisheries Service), got back the word that *"very heavy receipts of fresh swordfish were coming in from Canada."* The Bureau told him that in 1962 imports of fresh swordfish, practically all of it from Canada, totaled 3,622,000 pounds.

Imports in 1963 were racing well ahead of that figure. Senator Saltonstall was told that in June alone, 300,000 pounds of swordfish had come down from Nova Scotia and Cape Breton Island, an increase of 215,000 pounds for that month over the year before. In July, Canada's fresh swordfish imports into the United States rose to nearly 2,000,000 pounds for the month, an increase of three times what it had been for the same month in 1962. In discussing the matter with the State Department, Saltonstall was told that the Geneva Trade Agreements of 1948, which the United States had signed, stated clearly that tariffs were not to be exercised except against those commodities whose production was aided by grants or subsidies by the government of the producer. The question then was: Did Canada's fishing vessel construction subsidization program, then in full swing, constitute such a *"grant,"* or was it to be likened to the case of a governmentally subsidized merchant marine? If the latter was the case, then the hope of a protective tariff was dismal. Those persons most familiar with the workings of international trade law held out for little success. The less sophisticated continued to hope; but hope was all that came of it.

As with other fisheries, the Canadians appeared to have the upper hand. The United States fleet, without government construction aid, was in no position to be competitive. The Canadian government, in an attempt to build up its fisheries, was even subsidizing fishing vessel insurance. Insurance costs in the case of United States fishermen were becoming prohibitively high, so high in fact that accountants were counseling clients against entry into the fishing business on that basis alone.

In September, the Canadian Department of Fisheries' experimental ship, *Beiner* (117' long), was eased alongside a Newfoundland fish dock where her crew began unloading 293 fish for a total dressed weight of 58,662 pounds. (This averages out to about 200 pounds per fish.) *Beiner* had been out only twelve days. Jubilation was high until it was learned that the dockside price would be a disheartening eighteen cents a pound. An editorial article appearing in the *Canadian Fisherman* in November pessimistically appraised the price structure, stating that although the catch for 1963 was up over 300 percent from 1962, the revenues to the boats had only increased by a mere 10 percent. Within the same article, some individual trip landings were quoted: *Margaret M,* with 250 fish landed at Sydney; the 65-foot *Caress II* with 207 fish landed at Sambro. The writer claimed that there had been off-loaded on the docks at North Sydney around 3,000 individual swordfish during the previous August. Nevertheless, there was one optimistic remark attributed as coming from a Canadian official who stated he was *"confident that the low prices might be an advantage, in that a larger consumer market would be created, thus paying off in the future."*

Some boats on both sides of the border began dropping out, but the majority continued to fish. Even the research vessels kept at it, and from one of their trips comes to us a true story: While off Georges Bank in the fog, an American researcher's longline was being hauled back. Suddenly, looming out of the mist came the bow of another boat. Was she on a collision course with the research vessel? No! The other boat was busy hauling in gear, fish, yes, everything -- all of it belonging to the researcher. Dishonest? Certainly. But a sure way to beat the rising costs of longline fishing gear.

The Bureau of Commercial Fisheries, in making an estimate for the year 1963, claimed that *"there were no less than thirty-seven American and one hundred Canadian vessels [longliners] working the waters from the Delaware Capes north to the Grand Banks."*

Finesse in fishing technique had improved during the year, particularly as it regarded those water temperatures in which swordfish could be found. During the past spring, Captain Klimm, working the waters south of Hudson Canyon, had found a complete absence of fish despite sets totaling 960 hooks. The water at the time was a chilly 48 to 50 degrees. Two days later he got the same poor results -- only two tuna fish. The next day Klimm spoke to *Rush* and to the *Christine and Dan,* the latter with the Larson Brothers of Martha's Vineyard aboard. The *Rush* had nothing; the Larsons had taken only one tuna. Klimm, setting again, produced one swordfish, this time in waters which were slightly warmer, 52° F. Still hunting, he found a thermal front with the warm side at 61° F. He made a set on the warm side of the front and took six swordfish. The rest of the trip was spent hunting thermal fronts, and when he found one, even if the warmer side was only well into the high fifties, he took fish. Near the end of that particular trip, Klimm found blue sharks becoming a problem; they were moving shoreward with the warming waters, seeking the same warm thermal conditions as the swordfish. One haul in warmer water took two big-eye tuna, one swordfish, and an impressively annoying *"one hundred or more"* blue sharks -- the water temperature then being an even 60° F. To illustrate how conditions could change, Klimm's next trip started a lucky streak. The first set, no sharks and sixteen broadbills. Another two good nights followed. The score for that trip: sixty-two swordfish and fifteen tuna.

Veterans like Paulson and Klimm continued to refine and polish their techniques, more than ever certain that temperature barriers were the answer. They and some of the other highliners were at the time cooperating in the pelagic fish studies being done by WHOI, and this association enabled them to borrow some sophisticated thermographic instrumentation. With these devices it was possible to probe the sea beneath them down to 200 meters or more. Whenever stratification was discovered, particularly at the edge of the continental slope, acceptable levels of swordfish would usually be found. It was such improvements in methodology which kept the hook rate at economic levels. It is doubtful that boats at sea in the summer of 1963 could have yielded as high a catch if the more primitive techniques of 1962 had still been in use.

As the fall began and the season lengthened into early winter, the American fleet moved southward from Georges Bank. Some Canadians continued to fish successfully on Georges Bank through February. For the American fleet, things were becoming ever more crowded, particularly off Cape Hatteras. *Cap'n Bill III* on 18 January 1964 had in close sight at one time *Katherine C, Debbie and Jo-Ann, Gulf Stream, Excellent,* and *Wayne Laurin.* At that time many of the boats were taking between twenty and forty fish a night. Acceptable fishing, of course, but there were lots of things to iron out besides just the oceanographic questions of where swordfish could be found. Gear needed a lot of refinement as did safer methods with which to handle it. An example of the inherent dangers took place on 24 January 1964 when Al Daigle, a crewman on *Katherine C,* took a hook in two of his fingers, pinning them together. That was, though, nowhere near as bad as what happened a year or so later when a Rhode Island boat had a man plucked off the stern with a hook imbedded next to his jugular.[12]

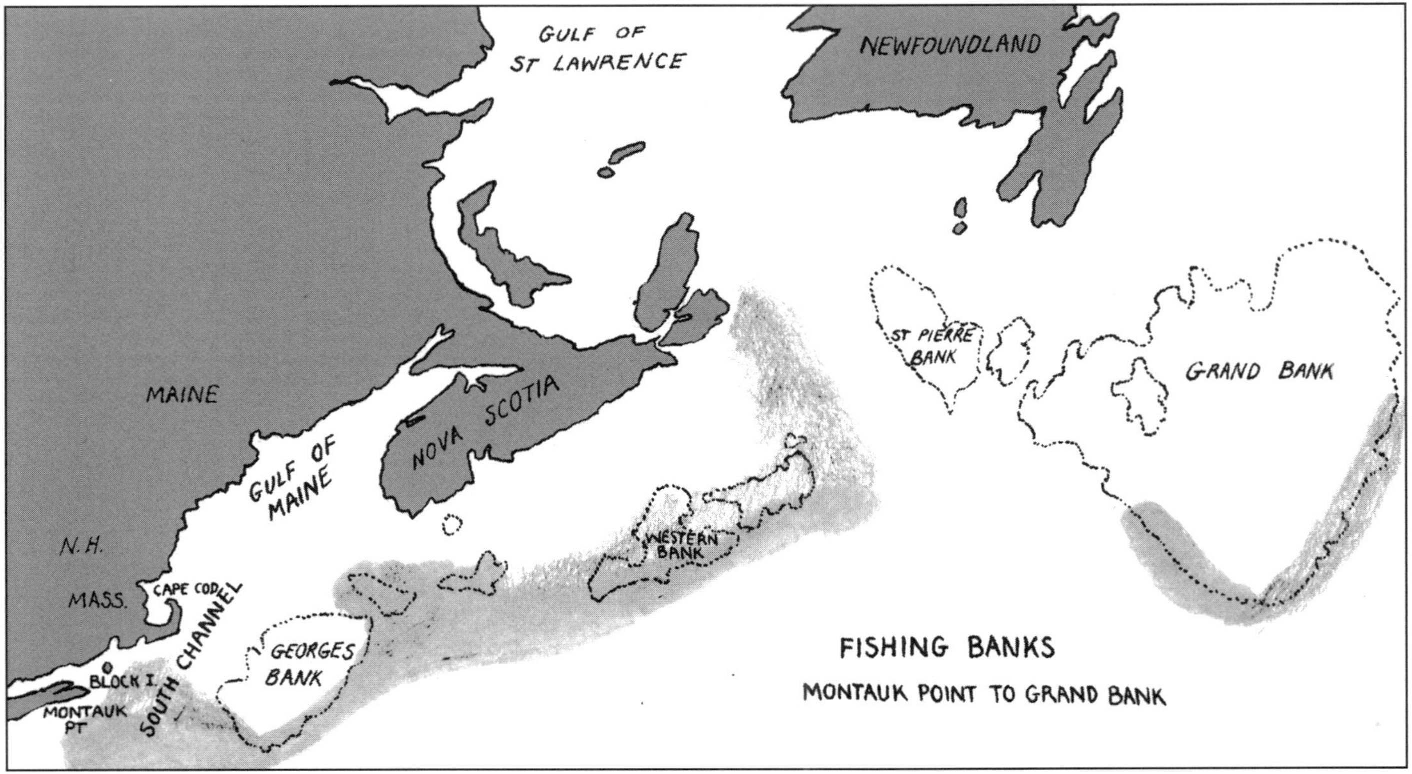

The swordfishing grounds from Long Island, New York, to the Grand Bank of Newfoundland are depicted circa 1960s and 1970s.

By the middle of March 1964, things appeared to be changing for the worse. Fish were thinning out and were becoming much harder to find. By May, fish again became plentiful, and it again began to look like a city out there. Both American and Canadian boats crowded into a small area of fish-producing water. This pocket of water, located along the west wall of the Gulf Stream, consisted of an eddy that had spun off from the Gulf Stream slightly o the northward of Oregon Inlet. As if things were not congested enough, more boats came out to join those fishing. Catches gradually lessened, yet there were still enough swordfish around to give experience to the newcomers, most of whom took between five and ten fish a night. Any tuna caught were thrown overboard as the price for them was only four cents a pound at the time. By the end of May, the fleet was fishing more to the north. *Katherine C* took seventeen fish on the night of the twenty-sixth. *Brandt* got twelve fish the same night with only 250 hooks out. On the twenty-ninth, *Katherine C* took twenty-three fish while *Harry Glynn* out of Rhode Island took fourteen. By June, the fleet had worked even farther to the north, in aggregate, fishing an area stretching from a little south of Hudson Canyon all the way to Hydrographer Canyon at the edge of Georges Bank. This is a considerable distance, yet because of the number of working boats, with each of them fishing anywhere from five to twenty miles of longline, the situation on the grounds was still crowded. Because of the gear density, veterans of the period say that the canyons at the deep-water edge of the slope became almost impossible to fish. When setting, inexperienced skippers would often cross the gear of others, not once, but sometimes two or three times, and in the fogs of early summer more than that. Vessels like *Gulf Stream* and *Cap'n Bill III* became seagoing Pied Pipers as the newcomers trailed them, hoping by staying in close proximity to match their success. As the summer of 1964 wore on, this fleet of American and Canadian vessels were heavily engaged in a predacious onslaught upon a fish population whose dynamics remained virtually unknown.

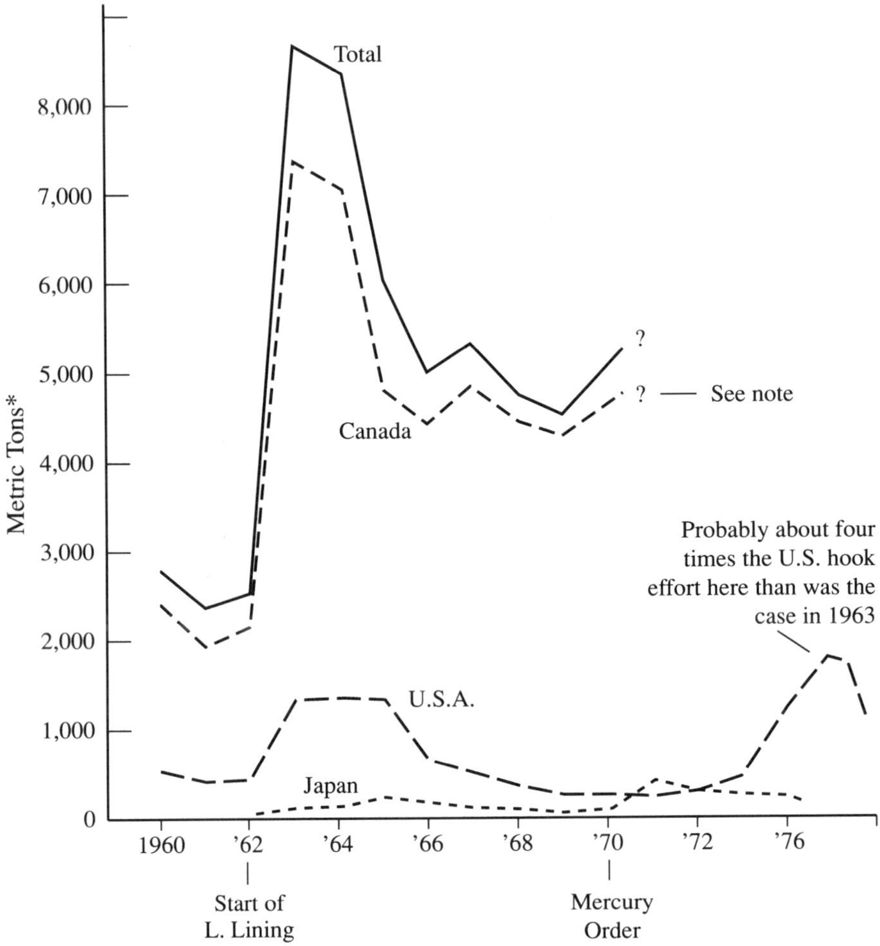

* A metric ton is equal to approximately 2,200 lbs.

Annual trends in the catch of swordfish in the Northwest Atlantic by Canada, the United States, and Japan, North of 25° N latitude and West of 50° W longitude. From Caddy Report Tech. 633, up to 1970. Thereafter United States data to September 1976, and Japanese data from 1971 through 1974 for their fishery.

Canada had a subversive fishery for swordfish from 1974 onward. Most of the fish taken by Canadian vessels were transferred at sea to United States fishing vessels and thus would reflect in the United States catch. From 1970 onward, the total United States catch as shown is to be viewed with some suspicion as a part of the landings were not reported. This was because of the mercury scare which brought about a federal ban against swordfish imports as well as those fish shipped interstate within the United States as well as within states working in conformity with the federal ban. However, the data which I used have been reinforced so as to roughly reflect this. United States catch taken in the Gulf of Mexico during the period 1971 to 1976 is not indicated due to questionability of tonnage landed; however, the amount is not thought to be significant.

From 1964 to 1967, swordfish production went down despite the number of boats engaged and the improvements in technology. Foreign imports continued to severely impact the price structure. The size of individual fish taken shrank dramatically. Fish under thirty-five pounds were now becoming common and were the reason for the coining of a new market term *"rat,"* the old term *"pup"* having been reserved for fish above thirty-five pounds but less than seventy-five pounds.

By mid-1968, the American longline fleet took a rapid decline. The Chilmark Fleet Company, organized in the mid-1960s specifically for the catching and marketing of swordfish, had already laid up its two boats. (The Chilmark Fleet Company is discussed in detail in the next section.)

While with the National Marine Fisheries Service, Martin Bartlett had been instrumental in developing a winter fishery in the Gulf of Mexico in the early 1960s. In 1969, a brief flurry of activity took place again in the Gulf; however, before success could be evaluated, the Food and Drug Administration (FDA) placed its market ban on all swordfish carrying in their flesh an excess of five-tenths parts per million (ppm) of mercury. In effect, the mercury ban killed the fishery all the way from the Mississippi deltas to the Grand Banks. The Canadian government soon followed with a ban of its own.

The swordfishery lay in a state of near complete dormancy until 1971 when a few fish began being landed at Massachusetts and Rhode Island ports. Massachusetts health authorities, which were by that time denying the FDA's position on mercury, allowed landing and marketing of swordfish in that state.[13] Emboldened by Massachusetts's policy and the now high price of swordfish, which was due in some part to the scarcity of domestic landings but in greater part to the absence of imports which had been totally prohibited by the federal government, American fishermen began re-entering the fishery. These fishermen discovered a stock which seemingly had multiplied since the hiatus of 1969 when the federal ban was first put in force. By 1973, the number of larger-class American longline vessels closely approached the numbers of 1963-1964. These boats fished to supply a consumer market which was apparently unconcerned with

FDA's still-vehement warnings regarding the dangers of consuming swordfish. Boats which were not unloading their catch in Massachusetts because of distance or other reasons were smuggling their fish to dealers in that state by truck or air freight. Some of those shipments were apprehended by the FDA, but most went through to Massachusetts without detection. Some Canadians began fishing again even though their catches were taboo in Canadian ports. These Canadians were finding ready markets by smuggling truckloads of fish across the United States border or by off-loading their catches to American vessels at sea. Those offloaded at sea to American vessels became reflective of the landings being reported in Massachusetts.[14]

The American fleet enjoyed a small production within the Gulf of Mexico during the winters of 1973 through 1975; but with a few rare exceptions, the catch there proved to be unprofitable to those engaged. Most of the swordfish caught between 1973 and 1975 were taken between June and November from latitudes northward of Cape Hatteras. This changed somewhat in the spring of 1976 with the development of a fishery initiated by the New Bedford fleet off Florida's east coast.

As had happened in the 1960s, once fishing efforts and landings increased, prices soon dropped. The pressure against prices -- and the fish population -- was aggravated by the influx of larger numbers of small vessels making one- or two-day trips from ports in Florida. The fishery overall, although declining in individual fishing boat profitability, remained relatively viable through the seasons of 1977 and 1978. However, by 1979, the catches had declined to such a degree that many boat owners moved their vessels into other fisheries.

FDA, unable to adequately support its mercury case on a scientific basis, and under legal opposition through a lawsuit conducted against the government by the law firm of Cadwalader, Wickersham, and Taft, had, for all practical purposes, abandoned the government's surveillance of landings by midsummer of 1978. The government's case for substantiating five-tenths ppm of mercury as a health hazard in saltwater fish was clearly invalid. FDA consequently raised its hazard level to one ppm of mercury. In reality, the continuance of any hazard level at all appears to have been a face-saving device for FDA. Despite the fact that some large swordfish had, upon analysis, historically contained well over a one ppm excess of mercury, FDA reportedly has conducted little, if any, monitoring surveillance since 1978.

In the immediate years following 1979, as documented by fishery scientists John J. Hoey and John G. Casey within an unpublished paper written for the Northeast Fisheries Center at Narragansett, RI, there was a steady decline in the number of larger class swordfish vessels (the traditional New England fleet). Starting the year before (1978), there had, however, been a noticeable increase in the number of smaller tonnage vessels operating for short trips out of southern ports, so if anything, the number of boats had actually increased. During this same time frame, radical modifications in the type of longline gear began to put in an appearance. While the New England and Canadian fleets had previously used longlines of hard laid nylon with gangions of similar material, the southern boats began using lighter and less visible monofilament, both for mainlines and for gangions. Chemical lights as attractants affixed to the gangions in proximity to the hooks soon came into universal use. Hooks were spaced at wider intervals (up to 100 meters apart) which materially reduced tangles. Catch rates per hook increased with the new gear. These improvements also allowed much longer sets. Some of the large boats began fishing as much as 40 miles of gear. The new type of gear, when coupled with improved designs in electronic fish finding equipment and pin-point navigational instrumentation, contributed to increases in per-hook catch rates. The National Marine Fisheries Service estimated at the time that efficiency per mile of gear had increased threefold. From a tonnage standpoint, landings were back up to earlier levels although weight per fish had decreased dramatically.

By the 1982 season, the New England fleet -- joined by converted steel shrimpers in the 65' -85' class -- had substantially enlarged the range being fished. Some vessels -- those with sufficient fuel capacity -- were fishing as far north and east as the Grand Banks. A few even ventured as far as the Flemish Cap, a bank which lies southward of Greenland. For long-ranging vessels, the Grand Banks produced the heaviest catches. According to logbooks examined by Hoey and Casey during 1982, 37% of the production of one high-line vessel (as tallied over a year-round fishing effort) had been harvested in proximity to the Grand Banks.

Taking the North Atlantic as a whole, the swordfish harvest for all nations during the early 1970s has been calculated as between 5,000 and 6,000 metric tons. By 1987, the harvest had climbed to 16,500 tons, and of that amount, the United States production was 3,636 tons for a dockside value of $27 million or about $3.40 a pound.

The winter of 1985 - 86 saw a total of forty-five U. S. longliners fishing distant tropical waters located northeastward of Puerto Rico. Others were working waters up to 1,000 miles eastward of the Lesser Antilles. By 1993, the largest of the vessels were fishing as far distant as the mid-Atlantic area north and a bit westward of the Azores Islands. Production for the distant waters fleet was high, but so was the fiscal outlay per voyage. Overhead for distant waters, inclusive of fuel, average gear losses, bait, ice, and of course food for the crew reached as high as $35,000 per trip. One of the costlier items had become bait which had tripled in price since the mid 1980s.

According to one account, 54% of the North Atlantic landings in 1987 were made up of fish weighing less than fifty pounds. Yet another study, also dealing with the production of fish taken throughout the North Atlantic basin for this general time frame, indicated that the adult population (meaning those fish of breeding age) had increased in numbers. With such numbers -- some slightly contradictory -- as a basis, U.S. fishing regulations were developed, aimed toward maintaining the adult population at a sustainable level. These regulations became effective in 1991. Among other things, they required log books to be kept by each vessel's skipper as well as the recording of swordfish receipts by wholesale dealers. There was also inaugurated the mandatory release of fish falling below 50 pounds and/or 31 inches in carcass length. The overall catch limit was first set at 4,560 tons. The fishery regulators realized, of course, that many of those fish released would be dead or dying when brought alongside the catching vessels, but this wastage was considered acceptable on the supposition that it would discourage fishing effort in those regions where there was an abundance of undersized fish in the catch. For those holding management responsibility, there was encouragement in that well over one-third of the U.S. catch was by now coming from the Grand Banks and that banks adjacent waters. This was an area where fish were averaging much larger than those caught elsewhere.

One major detraction from the otherwise optimistic management equations was that fish being caught in the mid-North Atlantic above 40° North were almost identical in average size to fish being caught by the Spanish longline fleet working the same latitude but to the eastward of the American fleet. This seemed to indicate an admixture of the stock in that northern latitude.

As will be stated in the Natural History section of this monograph, DNA testing has recently proved that the mid-ocean stocks above 40° North are one and the same population. Since American conservation regulations have no status with foreign governments, only international agreements -- a distant reality as of this writing -- can protect the breeding age swordfish population of the North Atlantic from over- exploitation.

PART I
APPENDIX

Swordfish Taken, 1991 - 1995

With the advent of vessel permitting, a program which was put into effect shortly before the 1991 regulations setting production limits, statistics became available on the numbers of U.S. boats which actively took swordfish (inclusive of those boats based in Puerto Rico).

Year	Number of fish *
1991	110,000
1992	91,000
1993	95,000
1994	104,800

* Numbers indicate fish retained as well as those discarded as being undersized or shark damaged)

Of the above numbers, 97.5% of the catch was taken by longliners. Gillnet boats made a relatively insignificant contribution of 1.5% of the catch. The effort of pair trawlers, as averaged over this period, was even less at 1% of the catch.

Year	Hook effort *	Number of boats which caught swordfish	Full-time boats fishing at least 5 months out of the year
1987	6.5	273	180
1988	7.0	337	216
1989	7.9	415	250
1990	7.5	362	209
1991	7.7	303	175
1992	9.0	295	137
1993	7.5	384	142
1994	10.0	458	167

* In millions of hooks.

Statistics published in the National Marine Fisheries Service's *Large Pelagic Logbook Newsletter, 1994* reveal catch success for U.S. vessels on a per area basis:

Area	Number of fish
Caribbean area	19,050
Gulf of Mexico	13,863
Florida East Coast	17,416
Southeast Coast, North of Florida	15,497
Mid-coastal, North of Hatteras	8,802
Georges Bank to Nova Scotian Banks	4,019
Grand Banks to NE of Azores	20,994
Mid-Atlantic - North Equatorial	1,752
Mid-Atlantic - Bermuda and East	3,456

A study conducted jointly by American, Canadian, Spanish, and Japanese fishery statisticians for 1992 and 1993 enumerated the North Atlantic catch for the vessels of each representative nation (all fishing methods). Figures are in metric tons.

	United States	Canada	Spain	Japan
1992	3,851	1,547	6,670	1,044
1993	4,191	2,234	6,594	No figures supplied

For 1994, the longline catch by U.S. vessels only (a statistic which included live and dead discards of undersized fish) was estimated at 3,872 tons, a figure which was later revised slightly upward.

The 1995 National Marine Fisheries Service (NMFS) preliminary report of the status of the swordfishery speculated that a decline of the fishery in *"recent years"* (as compared to the earlier years 1989 and 1990) was not reflective of a population decline but rather was a direct result of the establishment of landing limits.

Recent Tagging Results

Since 1991, the American observer program resulted in the placement of observers on many of the U.S. longliners. The result has been an increase in the scientific sampling of fish as well as an increase in the tagging of undersized fish which were released. Between 1991 and 1995, annual releases have averaged about 1200 *"live"* fish of which around 350 were tagged. Between 1991 and the end of 1994, a total of 54 of these fish were reported as being recaptured. One notable recovery from recent tagging was of a swordfish released in the central North Atlantic south of the Grand Banks; in a little over a year and a half, it traveled at least 1800 miles to reach the area off Cape Canaveral, Florida, where it was recaptured.

The American Swordfish Fleet

The year 1989 was the peak year for fleet size as measured by numbers of participating vessels. Numbers of boats has since declined; however, many of the vessels remaining are much larger than in the past due to the need to fish distant waters.

Seasonally, fishing effort varies depending on the region. This is a causative effect, both of fish availability and those weather conditions to be expected in distant waters. One factor which, according to Hoey, Cramer, and Rogers, accounted for the statistically enumerated reduced fishing effort on the Grand Banks following 1989 and into 1992 was a shift of some vessels to the Pacific tuna fishery. Some of those vessels later returned to the Atlantic swordfishery and their return accounted for the increase of landings which have been reported from the Caribbean and Grand Banks areas since 1993. By 1993, the Caribbean and Grand Banks areas were producing 47% of the total U.S. catch.

Fishing voyages as they took place for the years 1990 through 1994 in the southeast Atlantic coastal area were conducted by small boats (most of them less than 50 feet); their trips lasted between one and three days and produced an average of 1,000 to 2,000 pounds of swordfish. Caribbean trips made by larger vessels, many of which worked out of Puerto Rico, stayed out between 8 and 14 days with average landings of 5,000 to 10,000 pounds. Mid-ocean trips, whether well to the eastward of the Lesser Antilles or those made to the Grand Banks (as well as southeastward of the Grand Banks toward the Azores), averaged as long as twenty days, with landings running between 20,000 and 30,000 pounds. Wastage of undersized fish (meaning those which were dead or dying when released) was calculated annually by NMFS utilizing a 2-year average in the mid-1990s, wastage for each area being: Southeast Atlantic coast, 12,300 fish; Gulf of Mexico, 5,100 fish; Northeast, 4,500 fish; Grand Banks, 2,600 fish; Caribbean, 2,000 fish.

The ability to keep fish in good condition on mid-ocean trips was the result of improved handling techniques as well as the utilization of salt water ice making machines carried aboard the vessels.

Reports from observers indicate for the above time period that for released undersized fish, the live against dead rate varies widely according to the fishing area studied. The 1993 observations indicate that in the Caribbean and the Grand Banks the live releases of undersized fish which were apparently in a condition to survive ran around 50 per cent of the total released while along the Southeastern Atlantic such live and well fish made up only about 25 percent of the total. For the Gulf of Mexico and off New England, live and well releases were about one-third of the total. The percentage of landable-sized fish which were harvested for market was enumerated against the total caught as being 79% for the Grand Banks; 86% for the Caribbean area; 45% for the southeastern Atlantic coast; 32% for the Gulf of Mexico; 38% for the area off the New England coast. (There were no comparative figures given for the mid-Atlantic area.)

Catboat *Natalie* equipped for swordfishing. This vessel is not rigged with dories, so harpooned fish were brought directly aboard after tiring on a keg attached to the harpoon line. Early 1900s. Source: New Bedford Whaling Museum.

Harpooned swordfish coming aboard schooner *Natalie* showing fight; landed by tail sling. Source: New Bedford Whaling Museum.

A typical swordfisherman, schooner rig, circa 1920. Note harpooner on bowsprit and lookout man on masthead.
Source: Thigpen Photography.

Harpoon man on pulpit of schooner's bow about to "strike" a swordfish. Man with keg attached to harpoon line will toss the keg overboard as soon as the swordfish is harpooned. Keg will later be picked up by doryman who will fight the fish prior to bringing it alongside for lancing. Source: Thigpen Photography.

Harpooning a swordfish, probably off Georges Bank. (Circa 1920s or 1930s.)
Source: Thigpen Photography.

Felix and Florence Hickey out of St. Johns, Nfld. Rigged for harpooning. Circa 1960.
Vessels of this type were converted into longliners using tub trawl floating gear. Source: Martin Bartlett.

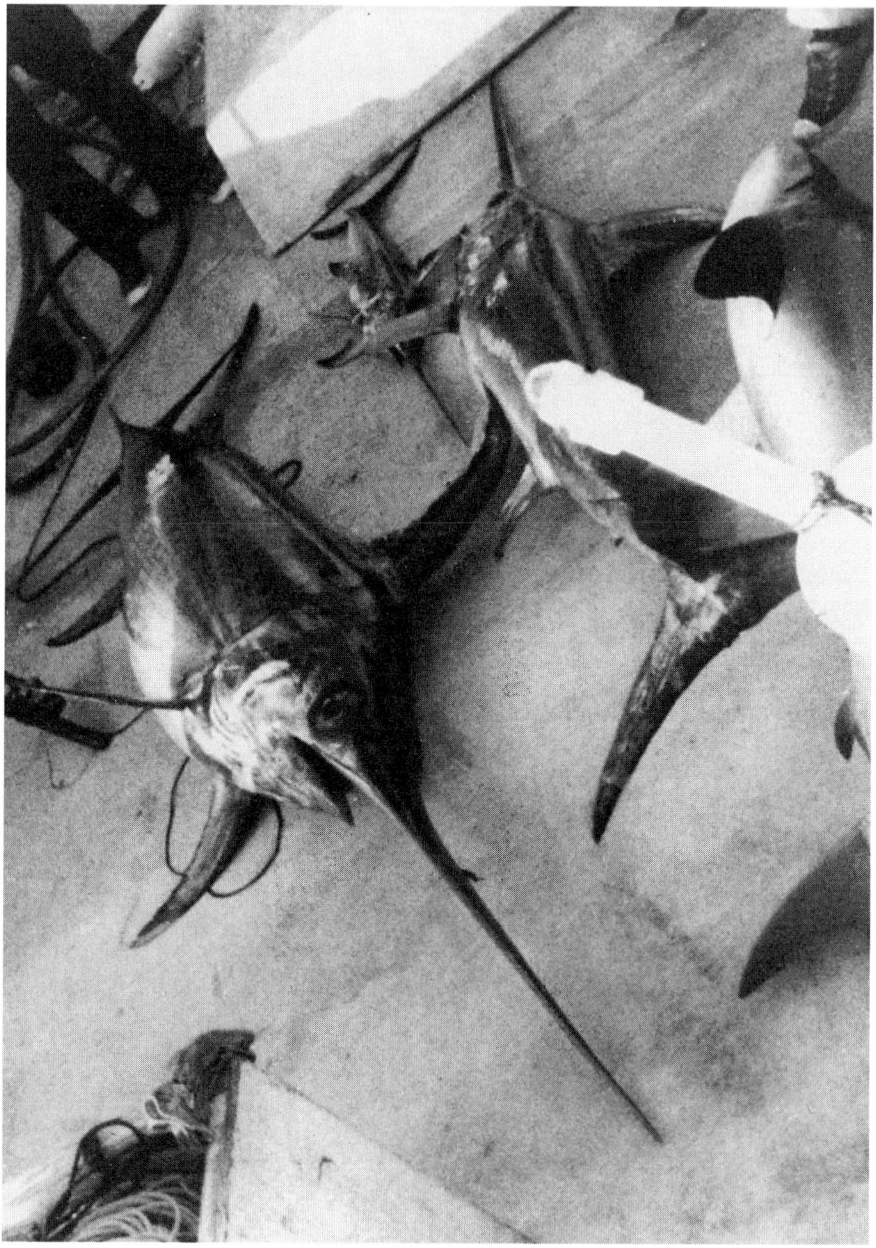

Three not untypical sizes of swordfish caught during a longline set.
Circa 1960s. Source: Martin Bartlett.

Gaffing a swordfish.
Longlining circa 1960s. Source: Martin Bartlett.

An unusual diet item. Eighteen seabirds (*Dovekies,* see fn 28) were taken from one swordfish's stomach. The lighter-colored items, also from stomach content, are squid. The fish was caught south of *Nantucket Lightship* on a longline set inside the 200 meter curve, circa 1960s. Source: Martin Bartlett.

Photo of a swordfish whose bill has pierced the bottom of a dory. Dory is being hoisted aboard the mother schooner with swordfish still attached. Circa 1920s or 1930s. Source: Thigpen Photography.

Deck view of longliner *Harry Glen* shown when she was equipped for tub trawls prior to a later installation of a longline reel. Circa 1960s. Source: Martin Bartlett.

The *Gulf Stream* out of Portland, Maine; skippered by Richard Paulson.
In September 1962, she took 119 swordfish on one of the earliest longline trips. Source: Martin Bartlett.

The *Chilmark Voyager* shown on sea trials following her launching in December of 1964. Source: Martin Bartlett.

Alvera C was a rather typical type of vessel which longlined for swordfish in the Gulf of Mexico during the 1960s.
Source: Martin Bartlett.

By the mid 1960s, most of the larger American longline vessels were equipped with reels of this sort. The line used during this period was hard-lay nylon. The reels varied in capacity from six to forty miles of line. By the late 1970s, monofiliment was being used instead of the nylon. Source: Thigpen Photography.

PART II

A CORPORATE PLUNGE INTO THINNING SWORDFISH WATERS 1964 - 1969

The story of the Chilmark Fleet Company is a unique episode within the economic matrix which made up the New England based longline swordfishery of the 1960s. Sheldon A. Dietz, a graduate of the Harvard Business School, would attempt to bring modern business thinking to the New England fishing industry. Although he failed in that attempt, the effort was innovative in many respects, and the brief story of that endeavor is more than worth the telling for the lessons it provides.

The Chilmark Fleet Company was conceived out of the swordfish enthusiasm which followed the introduction of the longline into the fishery in the early 1960s. The idea for the company came about because of the efforts of three fishermen from Martha's Vineyard working with a Boston-based alumnus of the Harvard Business School. The three fishermen created the basic concept; the businessman and his colleagues nurtured and promoted it. The fishermen were the Larson brothers, Bjarne, Dagbard, and Louis; the Harvard man was Sheldon A. Dietz.

The Larsons were relative newcomers to longlining although they were exceptionally accomplished at taking swordfish by harpoon. They had a strong heritage in this fishery as their father, Dan, also of Martha's Vineyard, had begun using his catboat *Cygnet* as a harpooner as early as 1908. As the boys grew up, they inherited as a natural course an interest in the fishery. Beginning with the 1945 harpoon season, the three began making seasonal trips to Georges Bank in their 57-footer the *Christine and Dan.*

By the mid-1950s, the Larsons had become a tradition at the Boston Fish Pier to which they regularly brought in the first swordfish landings of each year. By 1963, the Larsons could clearly see the advantage of using the longline in lieu of the harpoon and so the *Christine and Dan* was converted for longline gear. Encouraged with the catches they made by using the new method, but frustrated by the range and fish-carrying limitations of their 57-footer, the brothers began thinking of a newer and larger type vessel. Taking notice of the publicity surrounding the launching of a stern dragger, the *Narragansett,* built by Blount Marine Corporation of Rhode Island, they felt that she might well be the type of boat they had in mind. They subsequently bareboat chartered her for a field test which consisted of two longline trips. Impressed by the *Narragansett*'s sea kindliness and efficiency, the Larsons tried to buy her but failed to meet the cash asking price. They then decided on having Blount build them a vessel to similar specifications. With Blount's assist, the brothers redesigned the *Narragansett* concept, changing both the arrangement and with it, inevitably, the construction cost from that of the original prototype. Although feeling Blount's price was within reason, the Larsons realized that the cost would become a pressing commitment extending their debt capabilities to the limit. Obviously, some different route than individual financing would have to be arrived at.

With the above as the background, our scene opens, and Sheldon Dietz enters:

Dietz, a friend and summer neighbor of Louis Larson, was already enthusiastic over the brothers' venture and so he became an eager listener when Louie brought up the possibility of going public. Dietz had an established record of financial management, being among other things associated with a group called the Church Street Trust (*The Group*) which had made comfortable returns from various ventures, mostly to do with commercial real estate in the Boston area. With Dietz chairing, a meeting of *The Group* was called in early 1964, at which the idea of a private stock company to be formed solely for the harvesting and marketing of swordfish was discussed. *The Group* gave their go-ahead for Dietz to investigate further and report back. Dietz started off by putting together a hasty analysis of the longline fishery's history up to that time. This he reported back by memo to *The Group:*

February 12, 1964
Memo of conversation with Mr. McEvoy
Department of Fisheries [sic], *U. S. Department of Interior*

They [the Department of Fisheries, sic] *classify trawlers in three classes:*
150 gross tons and up = Large
50 to 150 gross tons = Medium
Under 50 gross tons = Small

Reports of U. S. Ships [fishing vessels making swordfish landings (longline gear)]

Gulf Stream, *170 gross tons -- a large trawler operating out of Portland*
1963 -- 11 trips, total catch 368,000 lbs.
Largest trip - 81,000 lbs.
2 trips - 54,000 lbs.
2 trips - 25,000 lbs.

1964 -- 2 trips, 24,000 lbs. And 18,000 lbs.

Cap'n Bill III, *113 gross tons*
1963 -- 12 trips, total catch 196,400 lbs.

Largest trip	*- 38,000 lbs.*	*1 trip*	*- 25,000 lbs.*
1 trip	*- 32,000 lbs.*	*1 trip*	*- 20,000 lbs.*
1 trip	*- 30,000 lbs.*	*rest - under 10,000 lbs.*	

.

Narragansett, [under bareboat to the Larsons], *113 gross tons*
1963 -- 2 trips, total catch 77,000 lbs. (42,000 and 35,000).

Christine and Dan, [Larson], *39 gross tons, a small trawler*
1963 -- 10 trips, total catch 130,4000 lbs.

Largest trip	*- 26,000 lbs.*	*1 trip*	*- 13,000 lbs.*
1 trip	*- 17,000 lbs.*	*1 trip*	*- 14,000 lbs.*

.

McEvoy claims that this market is controlled primarily by Canadians -- cheaper vessels, cheaper labor, cheaper everything. From 1954 to 1962 they exported between 4 and 5 million lbs. consistently. In 1962 their exports were only 3.4 million pounds; but in 1963, with the advent of the deep line [longline] *fishing, they exported 9.9 million pounds.*

U. S. total catch was 2 million in 1963, and McEvoy estimates probably an additional 200 to 300 thousand pounds were harpooned.

Duty on fresh swordfish is one cent per pound; on frozen swordfish, one-and-a-half cents.

Foreign vessels are not allowed to land catches in U.S. ports. Canadian swordfish are trucked from Nova Scotia down through Maine.

Swordfish are egg layers, and biologists feel there is not too much danger of thinning out the population via the deep line fishing. [Dietz may have misunderstood McEvoy on this particular point. Scientific evaluations of that period do not reflect any opinions as to catch prognosis.]

These trawlers also catch tuna on the same rigs as swordfish -- sometimes simultaneously.

Following his report, Dietz was given a go-ahead to investigate further. He began by garnering information from industry people as to the why's and wherefore's of marketing a fishery product, specifically swordfish. The questions put by Dietz were varied; however, he was consistent in basing his case on the same production volume which had been derived from the Larsons' claim that a vessel of adequate size and design, and engaged year-round in the fishery could average 20 trips a year, returning in the neighborhood of 40,000 pounds per trip in landings, receiving a projected price per pound estimated as averaging 40 cents. While Dietz was conducting his survey, so were some of his colleagues from *The Group*. Out of the latter came some cautionary input, namely, that the wholesale interests that had been contacted in New Bedford had developed serious doubts over the price stability of swordfish. This instability, it was felt, came partially from the habits of consumers who were unaccustomed to year-round supplies, as well as from the price competition being currently brought against the product by foreign imports. Canada was but one importer. A great deal of Japanese fish was showing up in the American markets, and the Japanese had developed a sophistication in cutting and packaging which would be a hard combination to beat. Bob Fairbairn, as one who had an interest within *The Group*, and who possessed good fishery industry contacts, became particularly cautious over the price. Fairbairn, in a letter to Donald Forté, *The Group*'s treasurer, reported that the Buzzard's Bay based proprietor of Captain Harris's Fish Market, had predicted prices to boats for the coming summer (1964) as going as low as 25 cents for large fish, and maybe 15 cents for *"pups."*[15] Already -- it was now February -- prices being paid to boats landing fares in Virginia were down to a dismal 35 cents. Apparently, the prospects for a stabile price structure as envisioned by the Larsons were lacking, unless, of course, Dietz the entrepreneur could improve the overall marketing structure. This is exactly what Dietz had said he intended to do; and on that premise, he hung the case for *The Group's* entry into the business. Dietz's theory that a new marketing approach was the answer was in part the factor which had encouraged the Larsons to join forces with *The Group.*

In the meantime, Bjarne, the senior of the Larsons, was engaged in final pricing negotiations with Blount Marine Corporation to build the vessel. In a letter dated February 17, Blount quoted a price of $140,695 for an 83-footer equipped and ready for sea (less actual longline gear); a second bid of $133,660 was given for a sister ship which was to have her keel laid shortly after the start of construction on the first vessel. By this time, *The Group*, generally undiscouraged by predictions of possible failure, had collectively decided to go ahead. A suggested organizational and operational format was worked out with the Larsons. It consisted of the following eight points:

1. *Compensation for the period commencing February 1, 1964, and running until boats are in water and operating will be made to the Larson Brothers at a rate of ...*[16] *per man per year.*

2. *Upon boats commencing to operate, the Larson Brothers will draw four (4) captain's shares and full crew shares with understanding that whenever any one of them are asked to leave the fishing to assist in any corporate capacity, that brother shall be reimbursed at the same rate as in paragraph 1.*

3. *The Larson Brothers shall have an option to buy up to one-third interest at original costs within 3-year period. In the purchase of the stock they shall be entitled to a 15% discount as consideration for their loss of income during the first two years. In others words, they can acquire ... stock for an outlay of ....*

4. *If the corporation is dissolved, the Larson Brothers shall have first option to purchase either or both of the boats at the original price less depreciation.*

5. *The Larson Brothers shall give the corporation six months notice, after three years of the launching of the boats, if they desire to discontinue their relationship with the corporation.*

6. *Sheldon Dietz and George A. Fink would desire a basis of compensation if they are successful in their distribution* [fish sales] *efforts after two years.*

7. *The possibility of divorcing the boats from the distributor corporation* [Chilmark Corporation] *will be considered to protect the Larsons' access to the boats in the event the corporation is dissolved.*

8. *The Boston price will determine the moneys due and payable to the ship on the day of arrival.*

By May, the final points of operation were worked out and developed into a contract between the Larsons and *The Group*. The venture was incorporated, and contracts for two vessels were let out to Blount. An addendum to this contract called for a third vessel, but only after the completion of the first two was the keel to be laid for the third. A review of the corporate correspondence for that period brings to light that even at that early juncture, *The Group* -- which would use the corporate handle of

Chilmark Fleet Company for the venture -- was aware of its investment vulnerability due to a total dependence upon the expertise of the Larsons. In fact, this dependence extended to all three of the brothers. This worry was focused by Bjarne's state of health which was far from good as he suffered from a chronic stomach ailment. To make matters worse, Bjarne provided the cohesive leadership for the brother trio. Since either Bjarne's death or retirement from active participation could result in severe management difficulties, the Chilmark Fleet Company (hereinafter to be referred to as *The Company)* took out key-man insurance on Bjarne, naming *The Company* as beneficiary. Tragic as it is to relate, this would turn out to be the smartest decision that had to date been made by the members of *The Group.* The worry was, of course, still quite real that the Larsons might just walk out leaving *The Company* very much in the role of holding the bag. Dietz, quite logically presumed that once a new and innovative marketing approach was arrived at, the Larsons would be locked in by the benefits forthcoming. Needless to say, all this was not the soundest foundation upon which to build a new business; yet the rosy picture of an ocean filled with broadbill swordfish and a latent army of consumers waiting in the wings to feast upon them overcame the not-so-rosy marketing forecasts.

Once *The Company* had been formed, then direct contact with Blount became a responsibility of Dietz rather than the Larsons, the Larsons having now taken on the role of port captains to *The Company*. Since neither Dietz, nor for that matter the Larsons, had experience with the building of modern steel fishing vessels, the shipyard part of the episode developed into one of considerable difficulties revolving around construction details.

It was during the construction phase that *The Company* received its first blow: A letter dated 15 July from Dietz to George Fink and Donald Forté (*Group* executives) announced the dismal news that Bjarne's illness had taken on new and serious dimensions, and it was doubtful as to whether he could ever go to sea again. Dietz wrote down the thought that Bjarne probably would be able to act in a capacity of port captain readying Chilmark vessels for sea once fishing started and also as a corporate marketing executive. Within the same letter, Dietz brought up plans to look into a permanent fleet port with packing and retail facilities to be operated by *The Company*. Point Judith was being considered as was New Bedford. From Dietz's viewpoint, both places had certain inherent problems: Point Judith, although relatively isolated, was thought to be a good prospect except for the risk inherent with a necessary relationship between *The Company* and the existing Point Judith Fishermen's Cooperative which was already located there. Dietz worried over an arrangement with the cooperative not working out, as it would have been necessary to share (at a price) certain of its facilities. New Bedford was deemed nearly ideal in most respects, although that port also had a risk factor. The fly in this ointment was that New Bedford's fish handlers were unionized. Dietz did not believe this to be a serious handicap due to the generally held opinion that the union's primary concerns were with the groundfish fleet. Port selection was to be left as an ongoing discussion to be largely dependent on the wishes of Bjarne, Louis, and Dagbard. Thought was still being given to that *"innovative marketing technique"* which Dietz had promised to come up with; but for now, the vessels' construction was the priority item on the agenda.

On August 17, Dietz, fresh back from a tour of Fulton Fish Market in New York City, reported to Forté his conclusions: Fulton Market, besides being larger, was much like Boston's Fish Pier -- *"a pattern of commission men, wholesalers to retail fish dealers, and hotel and restaurant suppliers."* Dietz's opinion of the merits of the New York markets' two primary swordfish wholesalers, Wilkinson and American Fish Products, was not as high as the one he formed for the firm of Caleb Haley and Company, which he initially selected to handle *The Company's* fish until such time that *The Company* developed its own marketing structure. Caleb Haley and Company had been picked for an assortment of reasons: it received high marks because of its market floorman, a son of the then present owner and who in Dietz's eyes was most importantly, *"a Columbia graduate."* One reading through *The Company's* files is hard put not to sense a communication gap of sorts between Dietz and the average operator in the fish markets of New York and New Bedford -- a communications' gap which may have inhibited his grasp of the way the industry normally did business. There is also a naiveté of sorts that can be gleaned from a report Dietz sent to Forté in which he suggests, *"in times of plenty they* (the dealers) *would have to take a minimum amount"* or that, *"We might run one or two trucks to deliver to them on* [our own] *route."* Dietz's expressed hope was to develop a chain of cooperating wholesalers at Boston, New York, Philadelphia, Baltimore, Washington, Chicago, Miami, San Francisco, and Seattle. At this early juncture, Dietz was seemingly unaware of the vagaries of fisheries in general and specifically what could only be an irregular production coming from just two (or even three) vessels operating within the oceanic environment of the northwestern Atlantic. As Dietz's inquiries and experience broadened, so did his realization (as displayed within *The Company's* growing correspondence) that marketing within the industry's established framework might be the best approach after all. It began to dawn on all those concerned that the greatest problem would be in catching the fish, getting them to market, and last but not least, collecting moneys due -- a conclusion hardly unique and most certainly in harmony with the thinking of countless other seafood producers from time immemorial.

While awaiting the launching of *The Company's* first vessel which would be named *Chilmark Sword*, another estimate of the coming production was made by Dietz in consultation with the Larsons. This one was developed to accompany an application for a government loan on the still undecided third vessel. The projection was derived from the Larsons' standing hypothesis of twenty successful trips over a period of a year.

PROJECTION

VESSEL: $150,000 equipped

STOCK: Gross Sales			*$200,000*
Fuel: 40,000 gals @ $ .13			
(2000 gal per trip x 20)	*5,200*		
Ice: 400 tons at $7/ ton			
(20 tons per trip x 20)	*2,800*		
		8,000	
			$192,000

Owner's Share (40%)			*$76,800*
Captain's Share (10% of Owner's)	*7,680*		
Larsons' Share (10% of Owner's)	*7,680*		
Boat's Share (3¼% of cost of vessel	*5,000*		
P & I Insurance ($330 per man x 8 men)	*2,640*		
Lumper Insurance	*100*		
Paint Boat and maintenance	*3,000*		
Gear Reserve			
(based on cost of new gear $7,000)	*3,000*		
Total expenses		*29,100*	
Amortization	*10,000*		
Interest	*9,000*		
		19,000	
PROFIT BEFORE TAXES BASED ON $200,000 STOCK			*$28,700*

Broken down, this projection came to a return before taxes of slightly more than 14% on a capitalization of $150,000 in a vessel.[17]

In retrospect, a fallacy in the hypothesis was the anticipated number of successful trips -- a figure far higher than what one might expect if realistically taking into account the factors of:

⇒ bad weather
⇒ failure to find fish consistently
⇒ breakdowns of the vessel and its equipment resulting in down-time both at sea and in port.

On September 23, 1964, amidst much fanfare and publicity, *The Company's* first vessel, the *Chilmark Sword* was launched. During the week following, she was taken to Martha's Vineyard for outfitting. By the 22nd of October, *Chilmark Sword*, under the command of Dagbard, was at sea. By the end of that month, Dagbard had already come in with a load of 236 average-sized swordfish. In November, *The Company's* second vessel, the *Chilmark Voyager*, came down the ways and by December 1 was also at sea under the command of Bjarne who, against doctor's orders had decided that port was no place for a fisherman. Fishing held good despite the normally poor fall and early winter weather. Trouble, though, soon cropped up over Dietz's marketing practices which resulted in delays over paying off the boats' crews. Fishing boat crews have an almost inbred suspicion of boat owners; in this case, their suspicion was aggravated by the abnormally long and arduous working hours of longlining in winter seas off Cape Hatteras. The share (payoff) system which Dietz had arranged with the crews was based on the *"average Boston price"* on the day of landings. That price was to be later refined after the fish were marketed by applying the actual prices *The Company* received -- less, of course, all marketing and transportation costs. The problem had its basis with the *Green Sheet,* an industry price reporting mail-out published by the government for major wholesale fish markets.[18] Within the *Green Sheet,* swordfish was consistently listed differently than the base price Dietz gave to the crews. What the men did not realize was that the Boston market price given in the *Green Sheet* often reflected only that day's early sales. Dealers report their prices to a government enumerator, and often these reports would be the dealer's attempts to keep prices high. Further, the wholesale prices given on the *Green Sheet* were grosses without deductions for sales commissions, nor did they take into account the consignor's packing and transportation costs. The average crewman on a fishing vessel, being only peripherally knowledgeable of the wholesale marketing system, rarely understands or appreciates the amount of overhead involved with the consignment of fish to a commission merchant. According to Louis Larson, as he explained in a letter to Dietz, the crews of *The Company's* vessels were no different in this respect. The slow delivery of checks to the crews, coupled with comparisons between price received and the *Green Sheet* produced unhappy rumblings in the fo'c'sles of the *Chilmarks*. Dietz had tried to educate the men as to the advantages of *The Company's* marketing. In retrospect, it would seem that the added expense of *The Company's* special sales efforts, along with the misunderstandings which resulted from it with the crews, was not worth any price advantage which could have resulted. Dietz himself realized this when within a management memo sent out to others of *The Group*, he expressed his belief that two more boats were needed to counterbalance *"Chilmark's* [The Company's] *extraordinary marketing and shore support efforts."* The advantage/ disadvantage ratio of *The Company's* marketing efforts would become increasingly more disadvantageous as the shore support of the vessels became a heavier load upon Dietz's time.

An experienced vessel owner, if asked for advice at that juncture in *The Company's* history, would have counseled that the best approach ownership can take is an efficient shoreside support of the boat(s), readying them for sea with minimal fuss and delays. A quality product landed from two such large and well skippered boats as those *The Company* possessed would in itself have provoked a competition among established dealers to purchase the fish. This, in turn, would by itself have given the price advantage that Dietz was seeking. It is also invariably the case that if an owner puts a maximum effort into the shoreside readying of his vessel(s) for sea, the result is good crew esprit. Dietz, and his by then two shoreside assistants, were so heavily committed to the selling aspects that they were prevented from giving the shoreside vessel support needed to produce smooth turn-arounds between trips. Normal procedure among the New England fishing fleet is that when a vessel comes in from fishing, the crew -- including the captain -- disappear ashore -- only to return when it is time to begin another trip. The extra two captain's shares which the Larsons received through the contractual agreements were thought of by Dietz as compensation to the Larsons for in-port boat management. The Larsons' interpretation, on the other hand -- harassed and worn out as they were by long, rough trips -- was somewhat different. The human factor is always present with such an arrangement; and a skipper, if not the owner of the vessel he commands, is not as eager to put in the port time he would otherwise. It appears that *The Group's* members as a whole may not have initially appreciated the strain laid upon the Larsons. The end result of these judgment differences was a widening gap in the relationship between the Larsons and *The Company's* functionaries.

At the time *Chilmark Sword* was launched, the fishery had begun taking on a downward trend. According to fishery technician Martin Bartlett, in an official Bureau of Commercial Fisheries evaluation, *"the 1963 longline fleet of 37 US vessels was reduced in 1964 down to 25."* Bartlett gave the reason for the decline as *"depressed markets, lighter catches, small fish, not to mention the competition for actual sea room to fish brought about mainly by the growing Canadian effort."*

To the Larsons, these new developments meant harder work made even harder by their conscientious efforts to deliver high quality landings. The dockside receipts made by the Chilmark Fleet during the severe conditions of that first winter's operations were fairly close in dollar volume, as averaged on a per-month basis, to what the Larsons had initially predicted. Expenses, though, were far higher than those they had predicted. The result was a net boat income falling well below original projections. This does not seem to have caused any great pessimism within *The Group* or even with the Larsons, perhaps because of the belief that the coming of spring would bring increased catches. That has always been the sort of optimism that sustains fisherman, and most *Group* members (especially Dietz) seemed to have shared that faith.

The Larsons' extraordinary endurance and their fishing abilities had been the primary reason for high production to date. Long trips and a willingness to fish in heavy weather was making up for the decline in the overall swordfish population.

The decline in the availability of swordfish (as well as tuna) soon became noticeable to others not engaged in the commercial fishery. The sport fishing interests were widely alarmed. Blame -- quite correctly as it would be proven -- was fixed against the longliners. American as well as foreign longliners, both in the Atlantic and in the Pacific fisheries, were extracting a heavy toll on the resource. Sportsfishermen from Acapulco, Mexico, to New England were beginning to turn militant. *Time Magazine,* January 28, 1966, told of confrontations. One sparked the sinking by rifle fire of Japanese tuna longline floats off Jamaica in the Caribbean. Another fracas was reported off the west coast of Mexico where a gun dual was narrowly averted between sportsfishermen and a Japanese mother ship. It was only the timely appearance of the Mexican Coast Guard that stopped bloodshed on that one. Perhaps the worst case of hostility toward longline fishermen occurred near Cabo San Lucas, Mexico. During the mid 1970s, the natives there related to the author the circumstances surrounding the 1960s demise of a modern Japanese longliner. (When viewed by the author, it was a broken rusty hulk ashore on the rocks.) The incident was brought about by American sportsfishermen who surreptitiously relocated a Japanese radio buoy from its normal position at the end of the vessel's longline. The new spot the sportsfishermen had picked for placement of the radio buoy was well inshore and atop a rocky headland. The Japanese skipper, steaming under nighttime conditions to retrieve his gear, was lured hard up on the rocky shore by the radio signal. Reportedly, the Japanese got the message, and this was the last serious incident along that part of the Mexican coast.

The Mexicans, as incensed as the Americans, soon began destroying every piece of longline gear they came across. One Japanese longline, overhauled by sportfishing vigilantes, *"was of 2,000 hooks having over 300 impaled sailfish thereon."* [19]

Most of the antagonism on the part of sportsfishermen was being directed toward foreign fishermen although Americans took their share of the blame as well. The publicity directed against oceanic longlining appeared in everything from the *New York Daily News* to *The Audubon Magazine,* and it created an awkward scenario for Dietz, who at the time was in the forefront of trying to promote swordfish before the American dining public. The adverse publicity, coupled with the influx of fresh Canadian fish as well as frozen imports coming from Japanese and Taiwanese sources, helped keep market prices down throughout the entirety of the 1964 season.

Dietz, addressing *The Group* by mail, made an analysis of *The Company's* 1964 -65 winter production in which he reported the length of trips and the stocks landed for both *Chilmark Sword* and *Chilmark Voyager.*

Vessel Chilmark Sword
First trip through eighth trip (22 October 1964 - 29 March 1965)

Trip No.	*Inclusive Dates*	*Stock Amounts*
1	*Oct 22 - Nov 3*	*$ 9,739.20*
2	*Nov 7 - Nov 21*	*16,856.91*
3	*Nov 28 - Dec 8*	*7,707.75*
4	*Dec 11 - Dec 20*	*7,779.03*
6a	*Jan 17 - Jan 25*	
6b	*Jan 27 - Feb 6*	*15,388.63**
7a	*Feb 15 - Feb 25*	
7b	*Feb 27 - Mar 3*	*7,940.50**
8	*Mar 12 - Mar 29*	*14,458.39*
8 Total Trips	***Total Stock***	***$79,870.41***

Average Stock per month: $79,870 / 4.75 mos. = $16,814.00
* *Totals for combined trips 6a, 6b and for trips 7a and 7b*

\- - - - - -

Vessel Chilmark Voyager
First trip through fifth trip (1 December 1964 - 13 March 1965)

Trip No.	*Inclusive Dates*	*Stock Amounts*
1	*Dec 1 - Dec 18*	*$ 9,073.94*
2	*Dec 28 - Jan 10*	*8,373.60*
3a	*Jan 17 - Jan 25*	
3b	*Jan 27 - Feb 7*	*20,128.53**
4	*Feb 13 - Feb 24*	*9,474.90*
5	*Mar 2 - Mar 13*	*18,018.90*
5 Total Trips	***Total Stock***	***$65,069.87***

Average Stock per month: $65,069 / 3.5 mos = $18,591.00
* *Total for combined trips 3a and 3b.*

For the 1966-67 season, both the *Chilmark Sword* and the *Chilmark Voyager* made catches which, when compared to the industry at-large, were in the highline category. Fishing was conducted off Cape Hatteras throughout the winter months and over the New England banks and some distance eastward during that summer. Production had been found to be best during the late fall and early winter; however, bad weather that winter had curtailed the fishing effort to an unprofitable degree. January of 1968 was relatively typical of what a handicap the weather could be. The boats only made one trip each and even then fishing time was merely a fractional part of the overall period spent at sea. After that experience, it was decided to take the boats south into the Gulf of Mexico, but the Gulf proved a disappointment as well.

By April, both boats were back in New Bedford. It was decided to take them out of service and lay them up at the D. N. Kelly Boatyard. This decision had been made some weeks before when Dietz, in consultation with *The Group,* had seen the handwriting on the wall. Things appeared doomed for failure with no realistic hope for improvement. That April, Dietz contacted a Puerto Rican development agency, proposing the use of both boats in a longline exploratory venture north and south of that island. He proposed that the base of operation would be at Mayagüez. The idea was that *The Company* would provide training aboard both vessels so that Puerto Ricans would eventually become the crews. He proposed both vessels being chartered for an aggregate of $96,000 for one year with *The Company* to be reimbursed *"the difference between receipts from the catch and operating expenses."* The boat gear would be provided by *The Company,* and a mainline length to accommodate 1500 hooks would be maintained at *The Company's* expense. During the training period, the captain, mate, and engineer for each vessel would be *Company* employees. Although some tentative interest was shown by the Puerto Rican interests, nothing came of the proposition.

Also that April, the senior of the Larson brothers, Bjarne, died of stomach cancer, and the contracts which the other two Larsons had with *The Company* were about to expire.[20] Both boats were put up for sale during which time an attempt was made by Dietz to capture a piece of the industrial marine chartering business. Two charters were negotiated, one with the Navy for a short term task of instrument calibration carried on offshore with submarines. The second charter, also of short duration, engaged one of the vessels for work on a geological survey in Long Island Sound. The prognosis made for obtaining sufficient charters to keep even one vessel going was dim. In fact, after those charters, both boats remained inactive at Kelly's Yard up to the time of their sales. The *Chilmark Voyager* was sold on May 9, 1969, for $145,300; the *Chilmark Sword* was sold on November 18, 1969, for $145,000.

The Chilmark Fleet Company, as of the end of business on February 28, 1965, had recorded a net operating loss of $27,984. In 1966, the loss carryover became $6,524, and in 1967, the loss was stated as being $39,151. In 1968, *The Company* reported a carryover loss of $45,421, and in the final year of business (1969), the carryover loss became $43,006. Up to the time *The Company* sold its two vessels, it had lost for *The Group* a total of $162,088.

Within correspondence with this author, Donald Forté, Chilmark's former treasurer, gave his personal opinion as to why the enterprise failed:

> *"In retrospect, I believe that the reason our venture was unprofitable was because we figured that we could fish following the thermocline for twelve months a year, but in truth, we were lucky to fish eight months. With so much capital tied up in the two vessels, it was a disaster having them lie idle for three or four months."*

Its swordfish venture was the first and last participation that the Church Street Trust *(The Group)* had with the seafood industry.

PART III

XIPHIAS GLADIUS, A NATURAL HISTORY

Xiphias gladius, known in popular vernacular as *broadbill* or as *swordfish,* is widely distributed in those open oceanic waters which have a surface temperature between 60° and 85° F. (Within this discussion, I will use the common term *"swordfish.")* The fish's temperature preference is probably around 70° F. Nevertheless, on the grounds off New England and Canada, these fish have been taken by ground line trawls when the fishing depths had temperatures as low as 41° F. It is equally at home whether on the surface or at great depths. (The deepest observation of this fish was off the Hawaiian Islands at a depth of over 730 meters.)

The Oceanography

To understand the distribution patterns of this fish, one must first examine the ocean current systems -- both warm and cold -- which affect the swordfish's environment within the Atlantic basin. In the northwestern Atlantic, from the latitudes of 8° North to those of the Newfoundland banks, the ocean currents are mainly horizontal in direction although some vertical movements (upwellings) do occur. The primary current systems which circulate off the east coast of the United States and Canada are the Gulf Stream and the Labrador Currents. These two currents are of particular significance to many oceanic (pelagic) creatures such as swordfish because of the influences of those currents on both feeding patterns and on population movements.

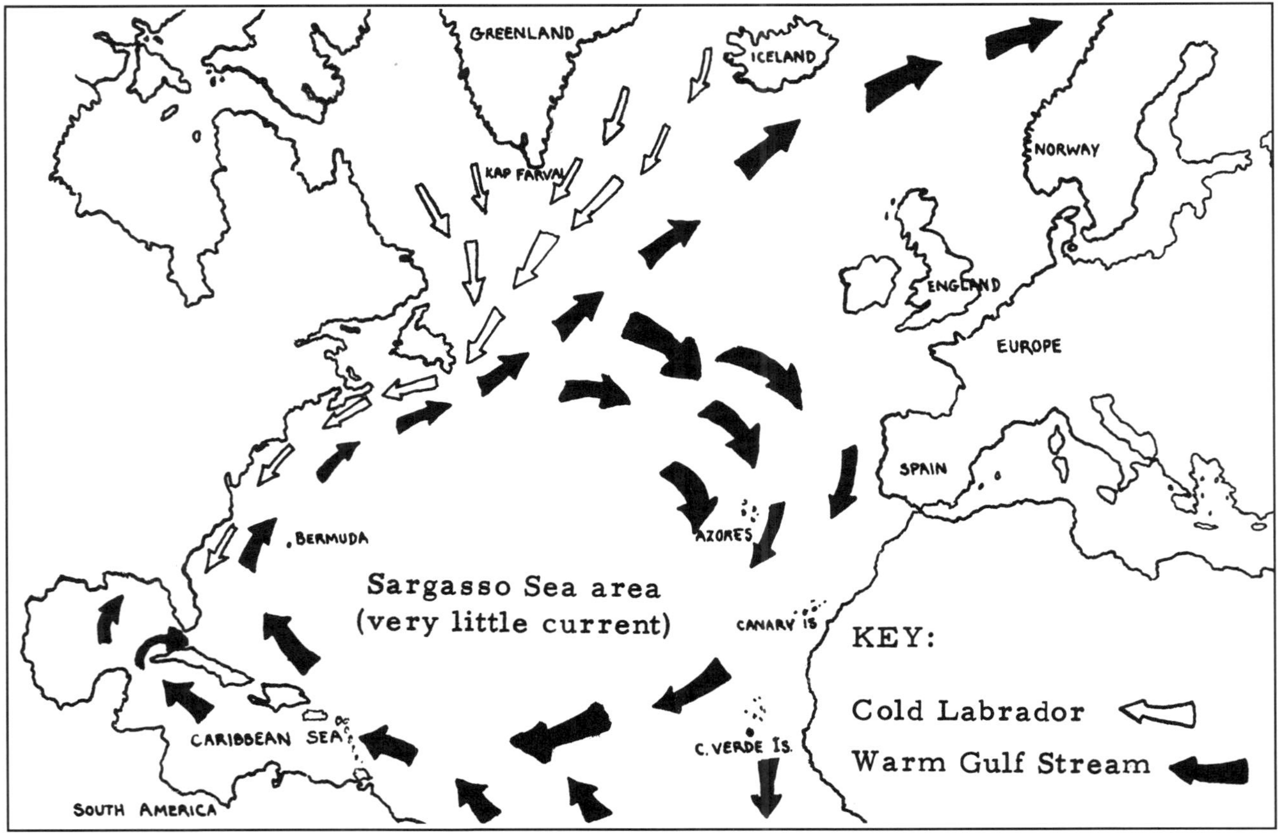

General circulatory system of the North Atlantic
Chartlet shows the major current patterns of the Gulf Stream and Labrador Currents.

The major summer and fall swordfish grounds which supported much of the United States's fishing efforts from the 1960s through the 1980s were at the offshore edges of the continental shelf from Cape Hatteras north to the Grand Banks off Newfoundland. Despite the wide difference in latitude, the waters on the outer shelf, from Cape Hatteras northward, do not vary appreciably in temperature during the late summer and early fall months. The warm Gulf Stream system passes at that season over or near to the offshore edge of the continental shelf. On the shelf is where the famous fishing banks -- the Grand Banks, the Canadian Western Banks, and Georges Bank -- are located. Inshore of the Gulf Stream is a cold water current. This current system is made up of nutrient-rich Arctic waters which flow out of the Labrador Sea and southward through the Denmark Straits. These Arctic waters move across the face of the Grand Banks off Newfoundland and then, joining two other contributory Arctic water masses, flows in a south and westerly direction over the Canadian Western Banks and New England's Georges Bank. These joined cold water masses are what is termed the Labrador Current. (Currents branching off from the Labrador Current are responsible for the cold waters of the Gulf of Maine.) After crossing over Georges Bank, the Labrador Current, now warmed to a marked degree, enters the so-called New York Bight where much of its southward movement detours to takc on a counter-clockwise circulatory pattern within that bight.[21] Thus, there are two primary current systems which affect the waters of the swordfish's environment within the northwestern Atlantic: the first being the warm Gulf Stream which flows up from the south, bathing the seaward side of the continental shelf; the second being the Labrador Current coming from the Arctic and which flows to the south and westward over the fishing banks of Canada and New England.

Although supporting appreciable sea life, the warm waters of the Gulf Stream are not nearly as rich as are the Arctic-originating waters which make up the Labrador Current. The fertility of the fishing banks of eastern Canada and New England is a direct result of the nutrient-laden Labrador Current.

The Paradox

The swordfish, known basically as a warm water fish, is found at certain times of thc ycar in cold (Labrador Current) waters. The reason for this is that although warm-loving by nature, the swordfish has a gargantuan appetite not easily satisfied in the tropic origin waters of the Gulf Stream. Consequently, when temperature conditions on the continental shelf reach a bearable level, the swordfish will move up onto the shelf where food is more abundant. Such favorable temperature conditions take place during the summer and fall when the water from the Gulf Stream becomes an increasingly dominant temperature factor at the offshore edges of the American and Canadian banks. At such times, the northern fishing banks become an accessible cornucopia for the foraging swordfish. Canadian and American studies made of both the harpoon and longline fisheries are in agreement that the female swordfish has the predominant presence on the banks, the male swordfish on average remaining within the Gulf Stream over deeper waters.

The Gulf Stream As a Path for Migration

It is known with certainty that within the northwestern Atlantic, the swordfish is an inhabitant of the Gulf Stream system. But what is not known with certainty is the extent and/or time interval during which any given segment of the swordfish population may follow the Gulf Stream's circulatory pattern. To examine the function of that current as a migratory route, it is useful to understand the dynamics of what causes that current.

The force which creates the movement of the Gulf Stream can be likened to the downstream rush of water from a hypothetical steep-sided estuary -- the water having initially been driven upstream either by a high tide or by strong wind. In the case of the Gulf Stream, we must mentally substitute the relatively confined sea known as the Caribbean for our hypothetical estuary example. An examination of a map of the western hemisphere will show that both the Gulf of Mexico and the Caribbean are largely encircled by land bodies with relatively small openings to the rest of the North Atlantic complex. The constant westward blowing winds between 9° N and 20°N (the northern trade winds) create a movement of water mass which builds in the partially landlocked Caribbean. This mass has its major outlet through the Straits of Yucatan. (The Straits of Yucatan are located between western Cuba and Mexico's Yucatan Peninsula.) At this constricted strait, a considerable current is developed which creates a water mass disgorging into the Gulf of Mexico. The mass then breaks off with part of it entering into the upper Gulf of Mexico while the remainder moves eastward into a constriction created by the Florida Keys and northern Cuba. The resultant current from that constriction becomes the Florida Current. At a position southward of Key West, the Florida Current begins angling northward, soon to be joined by the Antilles Current coming up from the Bahama Channel. (The Bahama Channel separates Cuba from the shallow shelf of the Bahama Banks lying to the north of the Bahama Channel.) When joined, the Florida Current and the Antilles Current become the Gulf Stream. The Gulf Stream, flowing as it does northward, is at first constricted laterally between Florida to the west and the Bahama Bank to the east, as well as vertically as it crosses the relatively shallow seabed between Florida and the Bahamas. It is here that it picks up its impressive speed -- at times exceeding 6 knots at its axis. Once northward of the Little Bahama Bank and therefore clear of both horizontal and vertical constrictions, the Gulf Stream begins to spread out, gradually losing much of its velocity. By the time it passes Cape Hatteras, its rate of flow has dropped by about thirty percent; and by the time it reaches the area just south of the Grand Banks, the current is down to about one-half knot. From Cape Hatteras northward, eddies of circular patterns break off from the Gulf Stream flow. These phenomena continue as far north and east as the Grand Banks. The eddies become environments wherein swordfish congregate. Part of the main water volume which made up the Gulf Stream -- the part that is pertinent to the scope of this discussion -- continues an eastward circulation to temper the climate of northern Europe while other parts of it slowly bend southward to off the Portuguese coast where it mixes with colder and richer waters which flow southward from the English Channel and the Irish Sea. These joined water masses continue southward to approximately 15° N where now reheated by the tropical sun, they are deflected toward the Caribbean by the westward driving trade winds, thus continuing the clockwise cycle of the Gulf Stream circulatory system.

The Swordfish Within the Atlantic Basin

Beginning in the 1950s, the Spanish and the Taiwanese began conducting viable longline fisheries for swordfish within the waters of the eastern Atlantic basin. The Spanish fisheries covered an area stretching from the Bay of Biscay southward toward the African coast. By the late 1950s, the Japanese were producing swordfish throughout the entire central Atlantic -- both north and south of the Equator. Starting with the early 1980s, the Spanish were targeting swordfish as far south as Dakar. There was also a fishery for swordfish conducted by the South Americans off the Argentines which soon reached northward to off the Brazilian coast.

No one knows with certainty if parts of the swordfish population which either summer on the outer shelf off New England's Georges Bank or on the Canadian Grand Banks leave in the fall to follow the easterly flow of the Gulf Stream system, later to become a part of the fisheries of North Africa. Those swordfish making such a sweeping migration, if escaping the fisheries of the eastern Atlantic, may next appear off the Antilles, continuing into the Caribbean and then passing through the Straits of Yucatan into the Florida Current. There, they may meet other fish coming in from the east via the Antilles Current. Moving northward in the Gulf Stream, these fish may eventually reappear on or adjacent to the banks of New England and the Canadian Maritimes. This is a possibility which seems likely, at least for part of the swordfish population. At the University of South Carolina, work was started in the mid-1990s aimed at a genealogical measurement for determining racial characteristics of swordfish. By using DNA technology, it became possible to determine whether there is more than one population within the Atlantic basin. These recent findings definitely point to a homogenous population within the entire North Atlantic. Those swordfish sampled south of the Equator and within the Mediterranean Sea had DNA characteristics which indicate that they were of a separate population. Prior to the

University of South Carolina studies, it was generally believed that the bulk, if not all of the swordfish population found in the western North Atlantic was indigenous to the northwestern quadrant of the Atlantic basin. In part, this may still hold true; but the facts presented by the South Carolina study tell us that some fish certainly take the circulatory North Atlantic route I have just described.

One long-held migratory hypothesis relates around a relatively short range of movement which theorizes that upon the approach of cold waters in the fall, most of the fish which are found adjacent to the Georges Bank of New England, as well as those found as far east as the Grand Banks, retreat into the Gulf Stream to travel southward against its flow, to winter in the Caribbean or its adjacent waters to the east of the Antilles, as well as in the Straits of Florida. Some of the fish following such a path may winter in the Gulf of Mexico although some theorize that the Gulf of Mexico population may be semi-permanent residents there. Yet another hypothesis is that with the approach of winter, some swordfish retreat into the Gulf Stream, cross it, and then travel along its eastern margin toward the south where they either enter the Caribbean through the Antilles chain, or bypassing that route, reenter the Gulf Stream system via the Antilles Current of the Bahama Channel. With the coming of spring, the bulk of the fish then move northward with the flow of the Gulf Stream. Another hypotheses is that some swordfish may migrate from the summer feeding grounds into the Gulf Stream axis, remaining in the Gulf Stream at a northern latitude until returning summer water temperatures again favor a return onto the continental shelf. At those high latitudes, the Gulf Stream may be rich enough in mid-water bio-mass to sustain an over-wintering swordfish population, thus allowing the fish to return to the adjacent Grand Banks the next summer in well-fleshed condition. This theory has credence since the Japanese as well as the Canadians have longlined this part of the ocean for some years and have found swordfish there well into the winter. However, since these latitudes are infamous for the North Atlantic's worst winter storms, winter fishing operations have never been extensive enough to definitively establish the existence of a sizable over-wintering swordfish population in that area.

It should be noted that the fishery which is conducted within the Florida Straits and in the Caribbean produces some fish through the summer months, indicating that at least a part of the winter population found there does not migrate northward during any given year.

In light of the latest finding, a most probable conclusion can be made that many swordfish do not migrate on a regular annual basis, rather returning to southern waters only after irregular periods of time -- semi-annually or perhaps even tri-annually. These migrations are probably triggered by the urge to spawn. This would help to explain the simultaneous presence of some swordfish within various parts of the northwestern Atlantic at all seasons of the year.

The females and the males are known to react quite differently to environmental conditions. This last is evidenced within (unpublished) catch statistics compiled some years ago by Woods Hole Oceanographic Institution and within Cuban statistics compiled by Guitart and Manday. These studies related respectively to landings in New England of swordfish which were taken from Cape Hatteras northward, and to landings of fish taken from off the northern coast of Cuba. In the New England fishery, the fish were taken by the longline method; off Cuba, they were taken on short-unit float lines. Each set of data included records of fish taken in both the summer and in the fall-winter seasons. The female-male ratio of fish taken in those latitudes north of Cape Hatteras was found to be in almost inverse proportions to the female-male ratio of fish taken in Cuban waters. For instance, in the fall-winter catch, the northern waters produced 89% females while at the same season of the year off Cuba, the catch was 88% male. Based upon a rather vague and unproved supposition that the population may be roughly 50-50 in sex ratio, one can make the conjecture that female fish, because of their heavier nutritional needs, are to be found in more northerly latitudes whenever conditions permit, returning south only upon the urge to spawn. It should be noted that in the area of Georges Bank off New England, both male and female fish are taken during the summer -- the females being found farther inshore on the shelf, the males being taken on the outer shelf within the warmer waters of the Gulf Stream.

~

While knowledge of the migrations of the swordfish may be scant, there are many known facts concerning the fish. The areas where the majority of the northwestern Atlantic population spawn is well recognized; and there is a general acceptance of how the larvae are disbursed within the northwestern Atlantic current system. As earlier stated, we now know the swordfish population of the entire northwestern Atlantic is genetically homogenous, being unrelated to the populations found within the South Atlantic or the Mediterranean. Canadian researchers have done extensive work resulting in our knowledge of the fish's rate of growth. Much is also known about the swordfish's feeding preferences although this, of course, varies from bank to bank and from latitude to latitude. Insofar as the summer grounds of the northwest Atlantic area are concerned, there is probably sufficient tagging information to feel fairly certain that the individual swordfish possesses some homing instincts. Also known is the probable depth range of the fish. These bits and pieces can help to assemble a partial profile, and since it all starts with an egg, it seems appropriate to discuss the fish's natural history, beginning with that egg.

Fecundity

The word fecundity, means *"quality or power of producing abundantly."* In the case of most open ocean egg-laying fish, fecundity probably should not be related beyond that point when the spawn is delivered into the water. This generalization definitely applies to the swordfish, as from the time of spawning, the fish's direct influence over its progeny ends.

That the swordfish is prolific is a certainty. According to one paper on the subject, one 176-pound fish had within its roe membranes 3 million eggs; another, a 440-pound fish was estimated to be carrying 6.2 million eggs. As is the case with most fish, the chance of a single egg hatching and the subsequent larva developing into a juvenile fish is infinitesimally small. Once released, a single swordfish egg (as well as a developing larva) becomes but one tiny unit in the huge and varied plankton bio-mass of the current system into which it is launched. It is upon this plankton bio-mass that small foraging sea life, including very young swordfish, depend for survival. Should the swordfish larva escape predation and develop into juvenile form, the odds improve, but they still weigh heavily against survival since while in the juvenile form it continues to be the prey of larger pelagic predators. Two of these predators (themselves becoming the prey of mature swordfish) are the dolphin fish (*Coryphaena hippurus* and *Coryphaena equiselis*) which are often found crammed to the gullet with immature billfish. As the juvenile swordfish grows into adulthood, it remains the prey for others. Captain Louis Larson, a well-known swordfish skipper who harpooned for swordfish beginning in the 1940s, and who, as a longliner, fished from Nova Scotia southward to the Yucatan Straits, believes that a major dietary item of many oceanic sharks is the swordfish. It is generally thought that the most successful of these shark predators is the mako shark (*Isurus oxyrinchus*). At sea, as in the jungles, the rule is eat and be eaten.

As late as the 1920s, the swordfish's spawning area in the northwestern Atlantic remained a mystery to American and Canadian scientists. The female fish taken in the old harpoon fishery which was conducted upon Georges Bank and the Canadian banks were usually barren of developed roe. Every year, only one or two egg-bearing females would be harpooned in these northern waters. In most fish species, it takes a considerable period between the time when roe begins to formulate and the time when it is *"ripe and running"*; thus it may be reasoned that a swordfish appearing on the American banks off New England or on the Canadian banks during the summer and mid-fall, and showing no noticeable roe development, must have considerable time to go before spawning. During the early history of the fishery, it was believed erroneously that swordfish did not commonly frequent areas south of Cape Hatteras and that they therefore must do their spawning at some mid-ocean place and at some time between mid-winter and spring.[22]

Beginning in the 1920s, a number of surveys were made by oceanographic and fishery research vessels engaged in sampling the mid-water bio-mass at various locations in the Northwestern Atlantic. Fish larvae were netted by the use of plankton nets towed from such vessels. When conducted in southern waters, these surveys took swordfish larvae. Following collection, the larvae were age-estimated. By computing the rate and set of the current within the respective sample area, a calculation would be made as to where the larvae would have been spawned. From these studies, it was determined that swordfish spawn in waters of not less than 75° F. Such temperatures are to be found during the summertime in northern latitudes, but Cape Hatteras seems to have been the northern limit for swordfish larvae captures, therefore indicating that spawning had to occur down-current (or considerably southward) of that point. Above the latitude of Cape Hatteras, juvenile fish were no doubt present; however, since they are quick little swimmers and thus difficult to catch in plankton nets, it remains problematical as to the location and density of juvenile-stage fish at any point in the Gulf Stream current system. From these studies, it became reasonably well established that in the northwestern Atlantic, swordfish spawn below 31° N latitude (approximately the same latitude as the Georgia-Florida line) and at greatest frequency between January and May, but with some spawning taking place well into the summer. The Fishery Research Board of Canada, from materials collected during eight cruises made between 1965 and 1972, offers some of the best information on larvae collections. The Canadian researchers took 119 swordfish larvae during their cruises. No larvae were collected anywhere within the southwestern or central Sargasso Sea which, set apart as it is from the general current pattern, is sometimes referred to as a vacuum within the Atlantic circulatory system. However, by way of minor in contradiction, the Japanese identified swordfish larvae from all over the central Atlantic basin where water temperatures were at least 75° F, but they did not gather any significant quantities in the northwestern Atlantic outside the areas of the main circulatory current patterns. When the Japanese findings are coupled with the information taken from the Canadian surveys of 1965 and 1972, they inform that the greatest densities of swordfish larvae are to be found in three regions:

* The Straits of Florida-Cape Hatteras continental slope region
* The Virgin-Leeward Islands region of the West Indies.
* East of the Antilles, south of the Sargasso Sea. This last is an area which is at present (1995) the operational locale of part of the American longline fleet which fishes there during the winter and early spring.[23]

From these spawning locations, it becomes determinable that the Gulf Stream system can be viewed as an indiscriminate distribution medium for young swordfish. The rapid growth rate which is thought to occur in both the larvae and the juvenile fish makes it probable that via the Gulf Stream, at least some young swordfish will spend their first summer feeding upon the shelves of the northern banks. Fish as small as five or six pounds in weight have on occasion been taken off New England and Canada. Other young fish appear to escape the current's pull to dally awhile in the warm waters off the West Indies, Cuba, or Florida, probably not arriving on the northern banks until the second or third summer of their lives.

~

Swordfish larva taken from the Straits of Florida, 12 mm long

Young of swordfish, 37 mm long. (Facsimile in Lutken's *Spolia Atlantica*)

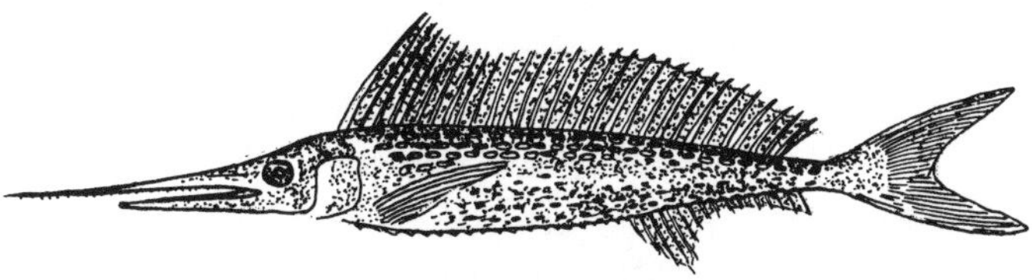

Young of swordfish, 12 inches long. (Facsimile in Cuvier and Valenciennes' *Histoire Naturelle de Poissons.)* Baby swordfish are miniature barb-wire fortresses, apparently playing a defensive role in their struggle for existence

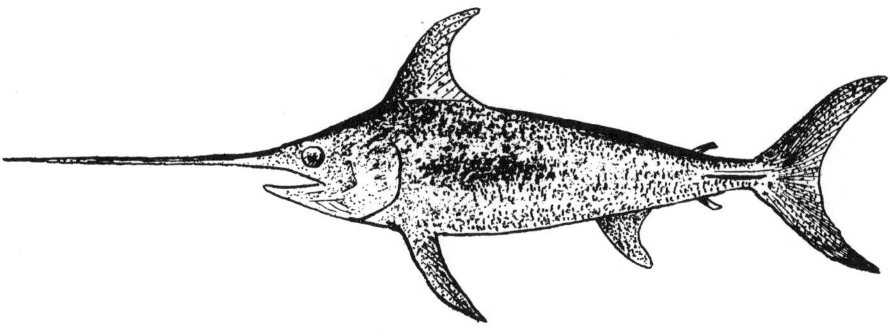

Adult swordfish are nature's submarines; they are U-boats and torpedoes combined, self-directed and self-propelled. (Illustration from G. Brown Goode, *The Fisheries and Fishing Industries of the United States.* Volume I.)

Shown on the image opposite are oceanographic conditions off U.S. and Canadian east coasts which are usual for November. This happens to be the situation for late November of 1976 and is from information supplied by satellite thermal infrared photography compiled by the National Oceanic and Atmospheric Administration. These images were used by many swordfishermen to help them plan their fishing strategy.

Note the warm slope extending north of Cape Hatteras. On this image, it is shown to be close to the continental shelf for at least 50 miles northward of the Cape -- thus affording a temperature environment suitable for swordfish and typical of that area during the late fall and early winter months. The warm eddies, shown on here and which are off New York and New England, are found in various locations and numbers (usually north of Latitude 36° N) throughout the year. It may be that from within these spin-off eddies, which become relatively stationary and which may last for days or even weeks, that swordfish larvae and small juveniles (spawned to the south during late winter through summer) may find their way to the shelves of the northern banks. While within the eddies the young fish would feed until large and mobile enough to migrate out into the main flow of the Gulf Stream, thus constituting the fishery's oceanic recruitment.

Initial arrival upon the coastal shelf by swordfish larvae or by young juveniles may initiate the "homing instinct" which tagging returns seem to indicate. (See Beckett's 1970 tagging report later in this chapter.)

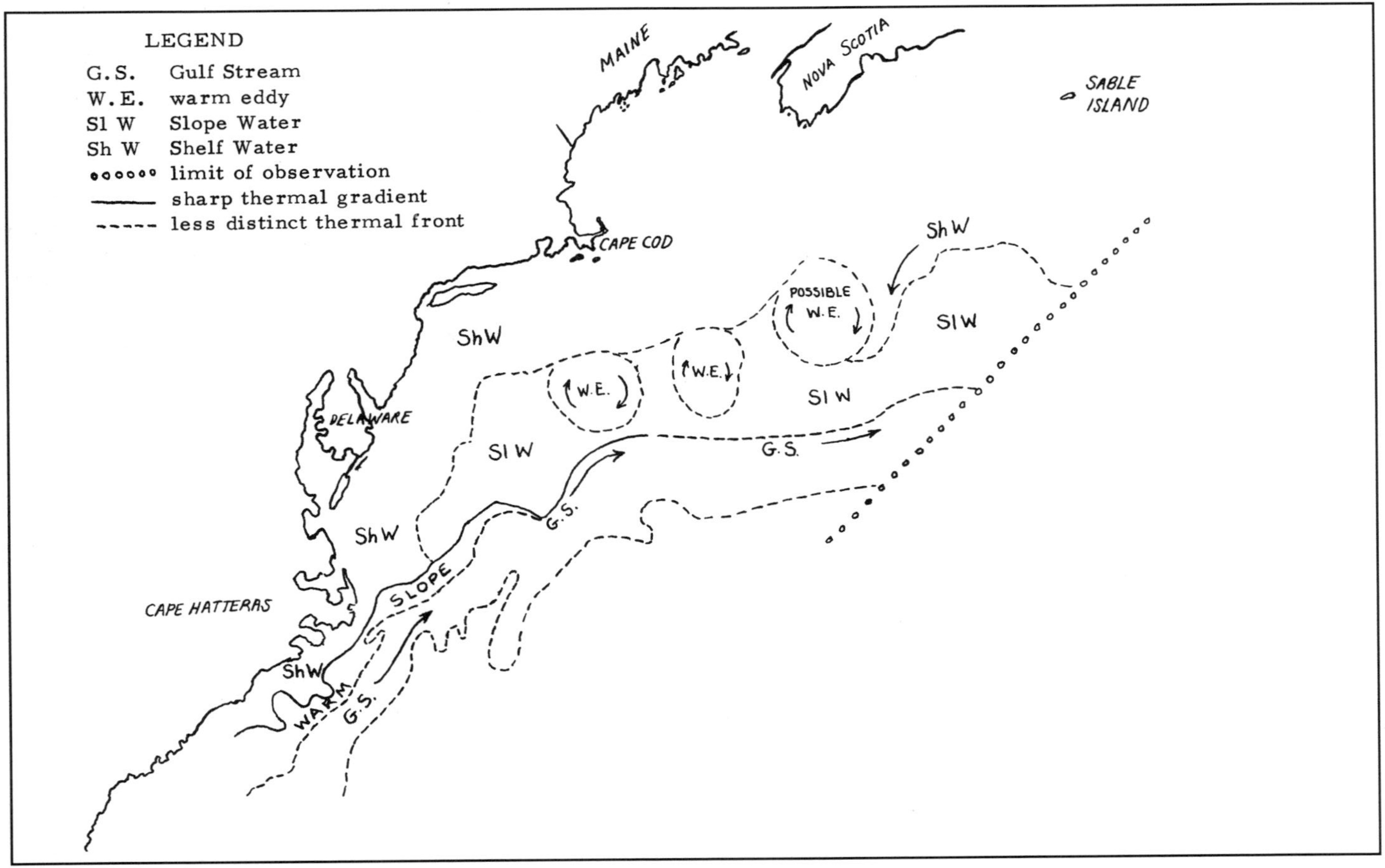
LEGEND
G.S. Gulf Stream
W.E. warm eddy
Sl W Slope Water
Sh W Shelf Water
limit of observation
sharp thermal gradient
less distinct thermal front
MAINE
NOVA SCOTIA
SABLE ISLAND
CAPE COD
DELAWARE
CAPE HATTERAS
ShW
SlW
POSSIBLE W.E.
W.E.
W.E.
SlW
G.S.
SlW
ShW
G.S.
SLOPE
ShW
ShW
WARM
G.S.

Feeding Habits and Food Sources

The feeding habits of a fish are important from the standpoint of how they influence that fish's general behavior since that behavior is related to harvesting by fishermen. In many fisheries, including the swordfishery, there is usually an observed difference between the food a fish seeks out in the wild and the bait it will readily take. In this regard, the swordfish exhibits a wide variance. For instance, in the Gulf of Mexico, mullet (*Mugil cephalus*) which the swordfish consume seasonally in great quantity as a natural feed, is a poor bait choice; the Atlantic mackerel (*Scomber scombrusa*), a fish foreign to waters of the Gulf of Mexico, is one of the better baits. On the northern banks, silver hake *(Merluccius bilinearis*) makes up a large part of the swordfish's natural diet, but it makes a poor bait fish. In the case of the hake, there may be an acceptable explanation since hake loses its sheen soon after it has been caught, thus not being reflective of light rays. Additionally, the hake has little structural rigidity and when on a hook, it looks more like a limp dish rag than a fish. Squid make mediocre bait in some areas, although in its live state, they seem to be a popular food item almost everywhere.

Nothing within the literature on the Atlantic swordfish resource indicates what a juvenile swordfish eats. Yabe, who studied the species in the Pacific, noted that as they reach the young juvenile stage, their diet includes small squid and fishes as well as free-floating crabs and shrimps. There is probably little reason to think that the generalized diet of juvenile swordfish inhabiting the Atlantic system would markedly differ from those studied in the Pacific.

As the swordfish grows to adult size, its feeding habits are of course closely related to the surrounding bio-mass. Studies performed by the Canadians on fish taken in their own longline fishery, that is in the waters from Georges Bank to the Grand Banks, indicates that the male fish (which universally run smaller than the female) tend to favor somewhat warmer waters which accordingly restrict them to the outer slope. The females, particularly the larger ones, seem to tolerate cooler waters and thus are located on the shelf where, while basking on the surface, they were historically taken in the harpoon fishery. Because their feeding locations differ at such times, the feed for both sex groups differs to a marked degree. This has been borne out through studies of stomach contents performed on harpooned fish taken on the shelf and on longlined fish taken on the outer slope. All sampling was done during the same season of the year and in the same latitudes. Swordfish at the edge of the shelf or on the outer slope, therefore, in deep water -- meaning most fish taken in the longline fishery -- fed on mid-water species which are most often referred to as part of the *"scattering layer."* Squids and lantern fish are such mid-water species as is the family *Barracudinas* (not to be confused with the barracudas, family *Sphyraenidae*).[24] The inhabitants of the scattering layer make vertical migrations upward in the water column upon the approach of darkness and downward at first morning light. Swordfish follow the scattering layer upon which they feed, descending upon the approach of daylight, while rising to within 50 to 100 feet of the surface at evening twilight. Since most of the creatures of the oceanic mid-waters migrate vertically, it at first proved difficult to determine with certainty at what time of day the swordfish did its feeding on this scattering layer. Swordfish have the ability to go even deeper than the usual daytime depths of the layer; however, in the longline fishery, they are rarely taken during the daytime (despite attempts to catch them at depths matching the location of the scattering layer during daylight hours).[25] It would appear from the evidence that swordfish only commence feeding when the layer starts migrating upward with the approach of darkness.

Stomach samplings indicate that when the swordfish comes into the shallower waters to be found on the continental shelf, it feeds both on the surface, in mid-water, as well as upon the bottom. One of the more interesting investigations indicates that swordfish have the ability to move from extremely deep water to the surface (and vice-versa) with very little time lag. Swift vertical movements appear to be the norm as a number of fish have been reported which had stomach contents containing feed taken from the bottom, the mid-water, and the surface layers, with no noticeable difference in the digestive deterioration of the three feed groupings. Telemetry studies performed by Carey and Robison not only confirm this ability of the swordfish to make wide-ranging and rapid vertical movements, but even of the tendency of some studied fish to hold stationary in the water column at considerable depths. One tracked fish remained at close to 600 meters for eight hours at one stretch; then within a period of an hour and a half, it ascended to within ten meters of the surface. The deep water stay for this particular swordfish occurred during the hours of maximum daylight; the ascent began shortly before evening twilight.

It is also deemed probable, as indicated from limited observations by Carey and Robison, that many swordfish move laterally on and off the shelf during their feeding cycles, coming up on the shallower shelf from the deep water outer slope when in pursuit of food. The southern longline fishery of the east coasts of Florida and Georgia -- which occurs outside the 200-meter contour curve -- indicates that swordfish at times rely completely on mid-water squid; however, at other times and in certain localized areas, fish caught well off the shelf showed evidence of feeding upon the coral ridges and peaks which are found along the inshore shelf at depths of between 55 and 120 meters. Martin Bartlett, who in the 1960s was a fishery gear specialist with Woods Hole Oceanographic Institution and later with the National Marine Fisheries Service, wrote of swordfish within the Gulf of Mexico having within their stomachs the partially digested bodies of *"parrot-beaked type fishes."* [26] A study made off Cuba during the winter months by Manday noted that 53% of the swordfish menu consisted of fish while 27% consisted of crustaceans -- many specimens from both of these groups usually being found in depths of not much over 70 meters. These observations seem to clearly indicate that both off Cuba and the Florida coasts the swordfish does much of its feeding within relatively shallow shelf waters (less than 100 meters).

The swordfish is undoubtedly an opportunist. Even octopae are occasionally catalogued in stomach content samples, although usually not more than one per stomach has been catalogued. Geography of course plays a major part in determining a swordfish's feeding preferences. As one proceeds northward along the coastal slope, the variety of species lessens. The waters south of Cape Hatteras offer a greater number of fish varieties than, say, do the more northerly Canadian Grand Banks; however, the overall bio-mass of the northern banks far exceeds that of southern waters.[27]

Between 1958 and 1965, researchers Scott and Tibbo examined the contents of 514 swordfish stomachs. These were of fish taken in both the harpoon and the longline fisheries; geographically, the samplings represented an area from southeastern Georges Bank, north and eastward to the Grand Banks. The stomach contents commonly contained: fishes of the family *Barracudinas*, the silver hake, red fish, squids of various species, Atlantic mackerel, and herring. Had this study been extended to southern waters, the results of course would have been different.

During the beginning years of the Gulf of Mexico's commercial swordfishery, Bartlett took an inventory of stomachs from swordfish that had been taken by longline both off the west coast of Florida and off the Mississippi delta. Off Florida, he reported finding:

squid	67%
butterfish	15%
tuna	5%
lantern fish	6%
other †	7%

† Included deep sea pomfret, lancet fish, and snake mackerel.

Stomach content inventories of fish taken off the Mississippi produced strikingly different findings:

mullet	35%
squid	15%
butterfish	16%
deep water viper fish	10%
other ††	15%

†† Included *"bumper"* fish, deep water shrimp, needlefish, the dolphin fish, ribbon fish, conger eel, moray eel, snipe eel, herring, catfish, blue runner, octopus, trigger fish, spade fish, wenchman, lobsterettes, silver hake, yellow fin tuna, *barracudinas*, and tile fish.

Whether the difference between the two areas in the samples studied was the result of the season or the location is conjectural. (The Florida sampling was in May; the Mississippi sampling was in February.) The point is that the swordfish's diet is without question well diversified.

The pièce de résistance for the swordfish appears to be bird life. A photograph taken while fishing an area south of *Nantucket Light Ship,* and which is reproduced herein, documents an event which occurred aboard the *Cap'n Bill III,* a longliner skippered by Captain Henry Klimm out of Woods Hole. Bartlett, who took the photograph, commented:

> *"On two consecutive sets, the stomach contents of the largest female* [swordfish] *caught were exclusively of the seabird known as the* Dovekie.[28] *Eighteen individual birds were found in the stomach of the first fish; twenty-two individual birds were noted in the second fish."*

Fish taken in other longline sets during that trip had only the usual mundane stomach contents of fish and squid.

The Swordfish Bill as a Tool

Bring up the subject of the swordfish to any gathering of interested persons, and sooner or later things will get around to a discussion of the use the fish makes of its bill (or sword) when gathering its food. The cutting edge of the bill is horizontal in plane thus indicating its use as a slashing weapon. A lot of the stomach contents of swordfish show cuts which would indicate the use of the bill for horizontal slashing, piercing wounds being rare. Scott and Tibbo give an excellent description of the use of the bill while feeding on various prey species:

> *"The swordfish appears to be more highly specialized for lateral slashing. Such a speculation would seem to be pointless unless directed toward a vertically oriented prey, or unless the swordfish slashes mainly while vertically oriented as when ascending or descending.*

"The food items were examined for evidence of slashing. The results are inconsistent; the percentage of stomachs showing evidence that the contents had been slashed vary from 10% in one sample to 53% in another...these figures show little except that the percent of slashing may be relatively high in a collection of stomachs. A stomach may have contained a slashed barracudina and a whole mackerel but would be considered as providing evidence of slashing.

"No consistent pattern was in evidence...Mackerel were often found to be cut into two or three pieces but equally often were whole. It does seem worthy of note that of the 24 herring in the stomach of one sample, not one showed evidence of slashing. It may not be sufficient, however, to look only for obvious cuts or decapitation as evidence of slashing. The use of X-rays indicated that although external evidence may be absent, the prey may still have a separated vertebral column. In one specimen, a gempylid, the break in the tissue above the fractured vertebral column took place after preservation in formalin.

"The fact that so many fishes removed from stomachs were slashed immediately behind the head (mackerel, white hake, and even the small lantern fishes) suggested that they were sought out and struck individually rather than as victims of a slashing attack on a school.

"The larger number of slashed Barracudinas *found in the stomachs of swordfish may be significant, particularly in the light of the suggestion by Houot and Rofen that these fishes swim in a vertical plane."*[29]

In analyzing some of the exotic items on the menu of the swordfish, i.e., octopae and the spiny lobsters as are found in southern latitudes, one could speculate that the fish may occasionally use its bill as a tool for dislodging prey from the bottom. Without using it to either toss or pierce, it would appear to be somewhat difficult for a swordfish to get at a prey which was crawling along the bottom. It most probably flips the spiny lobster or octopus off the bottom in order to position it for swallowing. A circumstance under which swordfish are known to use their bills -- and persistently -- has been experienced by fishermen of deep water tile fish (*Lopholatilus chamaeleonticeps*) when line fishing for that species was conducted along the east Florida coast beginning in the 1970s. On a number of occasions, tile fish which were boated had been repeatedly slashed; and in one reported case, a swordfish followed a hooked tile fish almost to the surface.[30]

The Swordfish Bill as a Weapon

One of the most discussed mannerisms of the swordfish is its tendency to turn on its attacker -- or for that matter, to attack without provocation. Are the attacks deliberate and made out of a natural aggressiveness, or are they made with a desire to defend itself?

Swordfish have attacked boats which range in size from dories to small ships. The literature recounts numerous incidents, some dating from the early 1800s. Related here is a small sampling of that literature.

- The yacht *Red Hot* was holed and sunk by a swordfish in 1871. This attack was reportedly not a result of provocation on the part of the yacht.

- Off Block Island on 15 June 1886, (an early time of year for the appearance there of swordfish) the schooner *Volunteer,* when two miles off Block Island, received a sharp shock which was likened to having *"struck a rock."* The crew, rushing to the vessel's side, saw a swordfish coming in for a second of what would become several attacks. Putting a boat over the side, as the schooner was at that time becalmed, the crew was able to harpoon the fish. When it was landed, it was discovered that the entire bill had been broken off, suggesting that some of its attacks may have been made without benefit of its bill.

At this juncture, it should be explained that the swordfish's bill is a formidable weapon indeed. It is bayonet-pointed and sharp on the edges. In its live state, the composition of the bone which makes up the bill is extremely tough and resilient. When thrust at an object, with the weight and force of the fish behind it, the bill has great penetrating power.

- A quotation from the New Orleans *Times Democrat* of 5 August 1883 is especially interesting since it appears to be the earliest reference to the presence of swordfish in the Gulf of Mexico. The news story relates that the schooner *Themus* of Boston, while journeying from Mobile, Alabama, to Tampico, Mexico, began to take on water. As soon as *Themus* entered Tampico, she was careened and her bottom was examined. It was determined that the bill of a swordfish had pierced the copper sheathing of the schooner. Not stopping there, it passed through the outer timbers of the hull which, altogether, were four inches in thickness. The bill then went through a vacant space of nine inches between the outer and inner timbers (or ceiling) and then penetrated some distance into the solid wood of the inner timbers.

- From the *Gloucester Times* of 22 June 1931, we read that *"leaking 400 strokes an hour, the schooner* Mary D'e Eon *had a hole made in her bow by a swordfish's thrust."*

The greatest danger from swordfish was not upon large vessels which in most cases could stand the punishment but rather it was upon dories launched from fishing vessels and upon their occupants when they were fighting a fish earlier harpooned from the mother vessel. For those not familiar with the harpooning procedure, the following is descriptive of the method as it was practiced off the New England banks:

A swordfish is harpooned from the bow (or pulpit) of the mother vessel. The harpoon line, attached to the iron (or dart) is allowed to pay out for the entirety of its length -- usually 400 to 900 feet of line. A float attached to the end of the line is next tossed overboard from the mother vessel. At this point, a (dory) is put over the side. It is manned by one or sometimes two crewmembers who then take the float aboard the dory and begin fighting the fish. After the fish has tired, it is brought alongside the dory, killed by a lance and secured by its tail to the dory's gunwale. The fish, the dory, and the dory's crew is then taken back aboard the mother vessel. There are scores of recorded instances which document injuries or near-miss injuries to dorymen -- the results of attacking swordfish. Whether the fish attacks in an attempt to rid itself of the pain of the dart is, of course, unknown. The fish may follow the direction from which it associates the pain until it reaches and rams the dory. Perhaps it attacks with the perceived knowledge that the dory represents the attacker. Some fishermen who have had the chilling experience of such an attack claim that they think the fish follows the harpoon line in an attempt to go after the source of its pain, some observers having seen an attacking fish follow the line's catenary as it arched away from the dory. On the other hand, there are instances which go beyond such a simple explanation.

- A doryman fighting a fish from the schooner *New Dawn* on Georges Bank in 1926 had his dory attacked with such force that it overturned it. The fish continued to attack over and over again, fortunately singling out the dory rather than the fisherman as the object of its attentions.

- A dory crewed by two men from the schooner *Pollyanna* out of Gloucester on 2 August 1933 was attacked by a fish that the two dorymen were trying to land. The fish cut [or smashed] their boat in half and then continued to slash (rather than thrust) at the two detached dory sections and at the men as well. Both men must have been quick and nimble as miraculously neither was injured.

- A fisheries scientist made a swordfish trip on the *Tubnico Pal* during the summer of 1959. While on that voyage, he was witness to two separate occurrences where swordfish holed the bottoms of dories.

Another explanation for the pugnacity of swordfish may be the brain damage that is brought about by a harpooned head shot which often causes the stricken fish to make erratic runs; many dories have been hit under such circumstances.

- The some-time ferocity of the swordfish is exhibited against craft other than fishing boats or their dories. Twice, submersibles have been attacked by swordfish. One instance involved the Woods Hole submersible *Alvin* which was attacked while exploring on the Blake Plateau off Georgia at a depth of about 700 meters. The swordfish rammed its sword into the gasket material of a viewing port. Fortunately, no damage was done to the submersible, but the swordfish, wedged tight, could not tear itself loose, despite its flailing efforts. The submersible was quickly brought to the surface, and the fish was taken aboard the mother ship, *Lulu, "where it was butchered and enjoyed by all hands."*

The question of whether an unprovoked swordfish would attack a free-swimming man has often been debated. To date, there has not been a reported case, but this is far from conclusive when one considers how rare it would be for someone to be swimming in waters frequented by a swordfish, particularly on the northern grounds where summer surface temperatures are usually between 59° F and to 63° F.

Growth Rates

There is no certainty as to the average growth rate of swordfish during the span of their lives, but growth appears to be quite rapid. During a billfish symposium held in Hawaii during August 1972, one contributor estimated the following growth rates based upon available information from various Pacific studies:

First year fish	9 pounds
Second year fish	33 pounds
Third year fish	88 pounds
Fourth year fish	150 pounds
Fifth year fish	240 pounds

Male swordfish run smaller in size than do the females. This may be due to morphology or it may be because the male fish, while in the northern latitudes of the North Atlantic, tend to inhabit the warmer, less nutrient rich zone of the outer slope. The females, on the other hand, feed in the colder waters which are found on the more prolific shelf. The smaller size of the male may therefore be a result of a lower caloric intake over its life span.

Equivalent weights for a given fork length* of Atlantic swordfish, and some tentative age equivalents (Beckett 1974).

Fork length (inches)	**Dressed wgt ** pounds**	**Tentative age estimated**
19.7	2.4	
23.6	4.2	
27.6	6.8	
31.5	10.4	
35.4	14.9	one year
39.4	20.7	
43.3	27.8	two years
47.2	36.4	
51.1	46.6	
55.1	58.7	
59.0	72.7	three years
63.0	88.8	
66.9	107.1	
70.9	127.9	four years
74.8	151.2	
78.7	177.2	
82.7	206.2	
86.6	238.1	five years
90.6	273.3	
94.5	311.9	six + years
98.4	353.9	
102.4	399.7	

* From caudal fork to tip of lower jaw.

** Decapitated just behind the eye, eviscerated, and with tail fin removed at the caudal notch. (Dressed weight is approximately 77% of round weight.)

Commercial Swordfish Landings

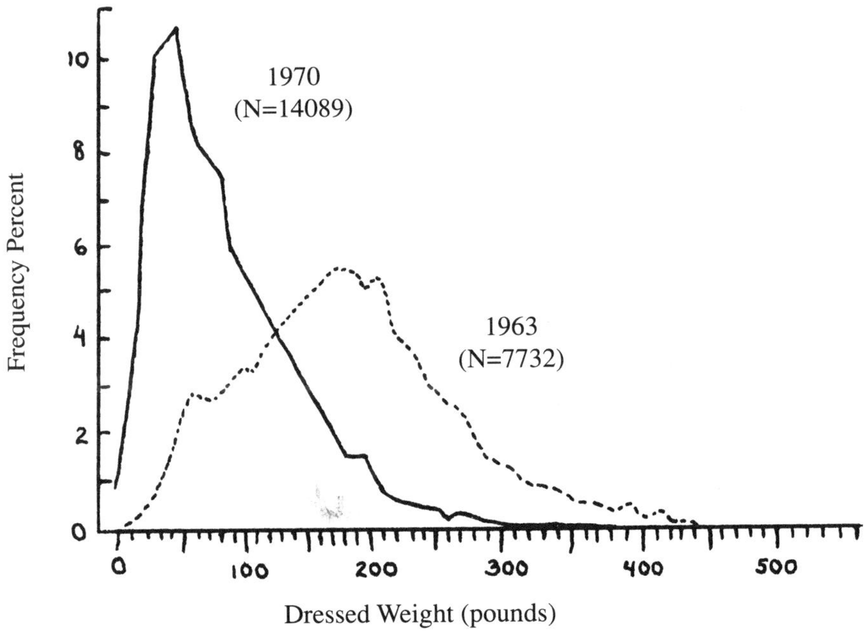

Size distribution of swordfish landed in Canada in 1970. (Dressed weight to live weight conversion factor 1.326). From *Biology of Swordfish in Northwestern Atlantic* by Beckett.

Homing Instinct

Tagging results suggest that some swordfish may possess a homing instinct. A report which was published by the Fishery Research Board of Canada in 1970 points toward this, but the small number of fish sampled make this more of a supposition than a certainty.

Swordfish tag returns (harpoon tagging) from Beckett's Report of 1970
Submitted to Fishery Research Board of Canada

Released D/Mo/Yr	Area	Size (lb)	Recovered D/Mo/Yr	Recovery Area	Size (lb)	Months out	Minimum Distance (miles)	Changed size (lb)
9/9/64	Georges	90 est	12/7/66	Georges	188 est	21	60	+98
7/6/66	Gulf Stream	70 est	10/7/69	Georges	156	37	128	+86
3/7/68	Georges	160 est	3/9/70	Stellwagon	212	26	178	+52
27/7/68	Sable	400 est	11/11/69	Sable	400+	16	7	0
27/7/68	Sable	350 est	2/10/69	Sable	590	15	6	+240
29/7/68	Browns	160 est	4/109/68	Browns	150 est	3	28	-10
13/7/70	Georges	120 est	20/9/70	Georges	140 est	2	59	+20
13/7/70	Georges	140 est	14/9/70	Georges	N/A	0	5	N/A
13/7/70	Georges	170 est	19/4/70	Georges	172	0	38	-20
13/7/70	Georges	150 est	11/9/70	Georges	185	2	83	+35
13/7/70	Georges	100 est	13/10/70	Georges	75 est	3	92	-25
13/7/70	Georges	180 est	27/7/70	Georges	228	0	31	+48
4/8/68	Georges	225 est	30/8/70	Georges	234	24	30	+9

Natural Enemies and Parasites

It can generally be stated that most sharks, particularly mako *(Isurus oxyrinchus)*, seem to prefer the same environmental conditions of temperature and food that swordfish require. This preference makes it not surprising that the major enemy of the swordfish is the shark group; of the sharks, the mako is probably the most destructive. Of all the oceanic sharks (including the great white), the mako is the best equipped to prey on swordfish. It is fast and highly maneuverable and has the type of dental gear suitable for the work. This is not to say that other common sharks such as the blue sharks (*Prionace glauca)* and oceanic white tips (*Carcharhinus longimanus*) do not take their share of swordfish, but considerable circumstantial evidence favors the mako as the more effective predator. The mako's feeding preference (live prey) as well as the ease with which it breaks off branch lines (gangions) make it less recoverable by longline than is the case with most other shark species. Because of that difficulty, it is hard to estimate the density of the mako population when measured against other families of sharks. The full impact of the mako or any other shark as a predator is therefore difficult to positively evaluate.

What usually surprises those who first are witness to a longline set which has taken a lot of shark is that only occasionally are free-swimming sharks seen. It follows, therefore, that the lack of visual sightings of free-swimming sharks is not indicative of low shark numbers in any given area.

To better understand the overall shark population of the Gulf Stream system, I would point to one particular swordfish longline retrieval made by this author in 1973 off the southern U.S. Atlantic coast (at a point off Georgia). A set of 600 hooks had been allowed to soak from 7 p.m. to the beginning of haul-back at 5 a.m. the following morning. The last part of the set was not taken aboard until noon as it had become necessary to search for sections which either broke off or were bitten off from the mainline. The soaking time for the gear was therefore much longer than average.

The catch for the set was as follows:

- 8 intact swordfish
- 8 shark-mutilated swordfish
- Close to 100 sharks of all types and descriptions, the majority consisting of the silky *(Carcharhinus floridanus)*, dusky *(Carcharhinus obscurus)*, hammerhead *(Sphyrna mokarran)* with some oceanic white tip *(Carcharhinus longimanus*) and tigers *(Galeocerdo*).

One mako was taken, it alone accounting for at least 100 meters of knotted snarls and well-twisted branch lines which were looped around partially deflated buoys, the deflation of which indicated a deep submersion. As the longline was hauled back during this not untypical set, it was observed that about thirty of the gangions had been bitten or broken off close to the hook. Around ten of those thirty gangions were badly snarled and twisted around the mainline. An educated guess would be that at least some of those thirty break-offs would have been caused by the powerful mako shark.[31] Despite their frenzy of activity after first being hooked, when brought in close to a vessel, makos are usually still very active. In marked contrast, other shark species (excepting the blue shark), when hauled close aboard, are most often either dead or in a state of torpid exhaustion.

A shark problem is also encountered by longliners in water adjacent to the New England banks. Shark varieties encountered there differ to a marked degree from those caught in southern waters. In the northern latitudes, blue sharks *(Prionace glauca*) are the most common. They can often be seen on the surface, but this is not always so. During the summer of 1976, while sampling for the presence of blue sharks prior to setting for swordfish at a location along the 500-meter contour offlying Georges Bank, the author launched six 1-hook floats with whole Atlantic mackerel baits suspended from gangions. Within thirty minutes, three of these floats were attached to blue sharks. The test was carried out three more times that same day at widely spaced locations with nearly identical results. Yet not once during that time was a shark to be seen on the surface.

Between 1961 and 1972, the National Marine Fisheries Service conducted longline survey cruises, making over four hundred longline sets from Cape Hatteras to Georges Bank. These sets yielded over 5,000 sharks, a catch rate of 6.3 sharks per one hundred hooks. The area fished was from Cape Hatteras to Georges Bank. The blue shark was the most common catch, making up 62% of the total sharks taken. The sandbar shark accounted for 10%; the dusky, 5%; the hammerhead, 5%; and the mako, 3%. The remainder of the catch consisted of less common shark species as well as tuna and billfish.

The longline method of fishing has had the certain result of reducing the numbers of sharks in proportion to the swordfish population. This is, I believe, true to its greatest extent in the waters south of Cape Hatteras where the shark population consists predominantly of hammerheads, white tips, duskies, and silkies. When these species were brought alongside in the early fishery, a sizable proportion of them were dead or nearly so from suffocation.[32] North of Cape Hatteras, the blue shark predominates, and longline destruction to sharks has apparently been far lower. This is because of the blue shark's agility which allows it, even when hooked, to move about sufficiently to prevent suffocation.[33] When cut loose, the blue shark usually swims off vigorously and, judging by the old hooks sometimes seen hanging from their mouths, they live to be re-caught many times. Prior to recent years when a market developed in the United States for shark fins and to a limited degree for shark meat, the blue shark population especially north of Cape Hatteras would therefore not have been noticeably reduced -- that is when compared proportionately to other shark species, especially to those species predominating south of Cape Hatteras.

During the height of the longline fishery in the 1960s and 1970s when sharks were considered a serious detriment to the swordfishery, fishermen made no consistent effort to destroy sharks taken on their longlines. Under normal fishing conditions where speed of retrieval of the longline is essential, gill lancing or other killing methods can be overly time-consuming. The usual and safest procedure was to cut the gangion, leaving the hook with the shark. The exception to this general rule was Captain Louis Larson who, in the summer of 1976, constructed a monstrous hypodermic needle out of a 7-foot length of pipe. He sharpened one end and attached the other end of the pipe to a hose that led to the vessel's air compression system. A valve was placed at the juncture of the pipe and the hose. If a shark was still lively when brought alongside, the point of the pipe *'needle'* was driven into the shark's belly cavity. The valve was turned on, resulting in a much-inflated and soon very dead shark.

While the larger sharks certainly take their toll on the swordfish population, the most macabre enemy of the swordfish is a little shark which grows only to about 18 inches. This small shark is reported to be of the genus *Etmopterus* and occasionally of the genus *Isistius,* but in all reports, it is given the common name *"cookie cutter shark."* It is a denizen of mid depths with an apparent range that does not extend north of 25° N. It is believed to attack tuna, whales, and billfish, inflicting small crescent-shaped bites. In examining individual swordfish, as many as half a dozen small bites attributed to the cookie cutter shark have been found on their flanks, dorsal areas, and most noticeably upon their fins. While fresh scars were occasionally seen on fish taken in the Cuban fishery, the destructive work of this small predator is more common and appears to be of more recent affliction on fish taken in the Gulf of Mexico, particularly those from the Mississippi delta westward to off the coast of Mexico. A supposition could be made that although this small shark's bites are not of themselves fatal to its victim, the effects of the bites (the release of small amounts of blood into the water) may be a trigger for attacks by larger sharks.

In addition to sharks, killer whales are capable of inflicting lethal damage to adult swordfish. Limited predation against young adult swordfish also takes place by large blue fin and yellow fin tuna.

While parasitism of various types is quite heavy against the swordfish population, there are three types of particular note: The sea lamprey, while not considered overly harmful, is thought by some observers to be the probable cause for swordfish breaching. (A term meaning to jump clear of the water.) One swordfish which was seen repeatedly leaping out of the water was shortly thereafter harpooned. When landed, it had a number of fresh, longitudinal scratches, suggesting that lamprey had just been shaken off.[34] Another is a strange and ugly parasite often found so firmly imbedded in the fleshy tissue of the swordfish that it cannot be pulled out by hand. The external surface (the part which is not imbedded) is leathery in composition. A fan-like appendage that looks much like an artist's brush protrudes from the end. This parasite is *Penella filosa*, a copepod. Stomach studies have disclosed that swordfish are often heavily infested with a round worm which sometimes numbers well over 100 per stomach. This worm is called *Contracaecum incurvum.* It seems to be universally present as it has been reported in examined stomachs over widely dispersed latitudes throughout the world. It does not appear to noticeably debilitate its swordfish host.

A Marketing Problem: Jellied Fish

A condition is sometimes encountered in swordfish meat after the fish has been refrigerated for a number of days. This condition affects the fish's flesh which is then referred to as being *'jellied.'* This condition may possibly be from a bacterial cause. It is included in this discussion as a known but relatively rare marketing problem. If indeed the condition is of bacterial origin, it appears to have no adverse affect on the live fish. Fish which have been harpooned but which were later found to be jellied, are reported to put up as good a fight against the harpoon iron as do other fish. The condition is not at first apparent. Fish which later become jellied, are seemingly in solid flesh when first dressed out and appear healthy in all other respects. That the condition is bacteriological in origin might be borne out by the observations that as an affected fish ages in storage, the jellied condition worsens. There are reported cases of fish being cut up and sold -- the meat appearing to be perfectly normal in appearance and texture. The next day, the housewife discovers that her once normal looking cut of swordfish is watery and jelly-like in appearance. The jellied fish cooks down rapidly and has little of the cellular consistency associated with an acceptable piece of swordfish; however, such a condition does not produce an objectionable odor and is apparently not harmful to the consumer.

If there is a pattern associated with jellied fish, it appears to be that the condition shows up in fish taken in the northern latitudes late in the season. It has been detected in fish taken as far south and west as Georges Bank but such reports are extremely rare.[35]

SUMMARIZATION

The quantitative information regarding the natural history of the broadbill swordfish (*Xiphias gladius*) indicates that:

The swordfish is solitary in its habits. Unlike many other pelagic fish such as the tunas, swordfish are not schooling fish.

The broadbill swordfish as it is found within the waters of the entire North Atlantic is closely connected for reasons of reproduction and migration to the current of the Gulf Stream and to the tropical water masses which contribute to that current. The swordfish's life cycle is completely pelagic. Appearing, as it does, to prefer warm water temperature, it remains, while in northern latitudes, near to or within the Gulf Stream or its eddies. The larger female fish venture upon the cold water banks when in search of food to satisfy their gargantuan appetites. Swordfish which are found north of Cape Hatteras concentrate along cold-warm water thermal barriers (or fronts). These thermal barriers occur along the outer edges of the northern fishing banks during the summer, fall, and early winter. In late winter, these barriers are more offshore over deeper water. In waters south of Cape Hatteras, current rips, often in sharp definition, are to be encountered along the western edge of the Gulf Stream throughout the year. South of Cape Hatteras, the rips or barriers do not possess the wide temperature differentials that usually exist in the more northern latitudes. A second oceanic thermal condition which seems to concentrate the swordfish is often found at considerable distances beyond the shelf and is a result of offshoot eddies of the Gulf Stream. These eddies create temperature barriers against which many types of pelagic fishes mass. Conditions of this type produce a sort of slowly moving oceanic supermarket in which both prey and predators are concentrated. Oceanic temperature barriers make the swordfish easily exploitable by fishermen.

Judging from recent DNA studies, it appears that swordfish taken within the fisheries of the western North Atlantic area are homogeneously related to swordfish populations found on the eastern side of the North Atlantic basin. They are not, however, homogeneously (DNA) related to swordfish populations in the South Atlantic or within the Mediterranean. As to what extent the swordfish populations found within the western North Atlantic mingle with the populations found within the eastern North Atlantic has not been determined. To date (1995), only one fish tagged in the Northwestern Atlantic has been recovered in the eastern fishery, but since to date there is no publicized program to alert foreign fishermen in the eastern Atlantic to American and Canadian tagging efforts, the lack of tag returns between the areas is easily understood. There is believed to be a considerable intermingling of swordfish taken by longliners of the American and Spanish fleets when the two fleets are operating in mid-ocean areas northwest of the Azores Islands. The American longline fleet seasonally fishes that area, generally working to the west of 30° W longitude while the Spanish longline fleet generally works to the east of 30° W longitude. A cooperative census conducted by American and Spanish fishery scientists has recently determined that fish taken at the same time of the year by both fleets in this area are closely identified in size composition. This could indicate that this region is the junction point at which western hemisphere swordfish (of specific age classes) begin a migratory swing which would take them south to off Africa and from there onward in a westward path to the Antilles and into the Caribbean. The percentage of the swordfish population making such a migratory journey during any one year remains an unknown.

From longline catch records of the swordfishery off Cuba (assembled by Cuban scientists in the 1950s) and which were later correlated against U.S. and Canadian records of fish taken north of Cape Hatteras to the Nova Scotian shelf during the 1950s and 1960s, it was determined that there is at certain times of the year a marked sex differential between the southern and northern populations found in the Northwestern Atlantic.

What can be said with certainty is that there probably is no regularly spaced migration with a segment of the swordfish population vacating one latitude for another or one hemisphere region of the North Atlantic for another. Proof of this is that within the fishery conducted south of 30° N latitude, swordfish are taken year-round which would not, of course, occur if the area was totally vacated through migration. Correspondingly, in waters north of 40° N latitude along the west wall of the Gulf Stream and continuing eastward to the Grand Banks, longliners take fish well into the winter.

Spawning seems to occur in southern waters in temperatures of around 75° F. For fish not urged on by the spawning urge, north-south and/or south-north movements probably are governed primarily by temperatures and the consequent availability of food. Spawn is almost never found in fish north of Cape Hatteras. This may be because the biological triggers which cause roe to develop are activated only when the fish is within a certain water temperature. There may be a sort of time-clock reaction which urges the individual fish to move southward for water temperatures more suitable for the development of spawn.

Predation is undoubtedly heavy against the swordfish, particularly while the fish is in its larval and juvenile stages. As an adult, it falls victim in unknown numbers to the shark of which the mako is probably the greatest danger. The swordfish population, dependent as it is for feeding purposes upon narrow thermal barriers, is especially vulnerable to shark attack since sharks are attracted to those same thermal conditions. There is strong circumstantial evidence which suggests that over the last three decades, the numbers of sharks has decreased in ratio to the swordfish. If this is so, then the predatory effects of sharks against the swordfish resource may have been lessened to a considerable degree.

As a resource, the swordfish's habit of feeding along oceanic thermal barriers makes it an easy target for fishermen. Swordfish concentrations can also be found on the offshore slopes of continental shelves as well as in areas near sea mounts and certain other oceanographic phenomena which tend to create vertical mixings of deep water layers with surface water layers.

EXPLANATORY NOTES

All Sections

1 Equipping a vessel for the harpoon fishery required the installation of a harpoon stand projecting out from the bow and lookout cages high on the vessel's forward mast.

2 The New York bight is described as an area encompassed to the north by Cape Cod and to the south by the Delmarva Peninsula. It is bounded offshore by the 200-meter curve.

3 Charles Dana Gibson, "Victim or Participant? Allied Fishing Fleets and U-boat Attacks in World War I and II," *Northern Mariner.* Ottawa, Canada: Canadian Nautical Research Society, Vol. I, 4, October 1991.

4 Captain Wagner's theory that fish were to be found off Cape Hatteras during the month of January was sound, and as we will see, a fishery of considerable magnitude did develop during the 1960's seaward of Cape Hatteras. However, in that area, as is the case in warm southern waters generally, swordfish rarely appear on the surface; therefore, a harpooner would meet little success even if fish were in plentiful supply.

5 They, of course, knew of fish farther to the eastward, but the grounds off New England were the extent of the cruising range for most of the boats operating from Massachusetts, Rhode Island, New York, and New Jersey.

6 In 1984, the International Court of Justice established a permanent fishery boundary between the United States and Canada. This boundary line gave the Northeast corner of Georges Bank and the Northeast Channel to Canada. See Donna R. Christie, "The Georges Bank/ Gulf of Maine Boundary Dispute Between the United States and Canada," within *Georges Bank.* Cambridge, MA: MIT Press, 1987, pp 469-73.

---

7 For a comprehensive discussion of wartime losses of fishing vessels during both World War I and World War II, see Gibson, *Northern Mariner,* October 1991.

8 It was generally believed at the time by fishery biologists and fishermen of long experience that the swordfish of Georges Bank were a different class of fish from those of the Canadian banks to the north and eastward.

9 A porbeagle shark (or mackerel shark), like the mako shark, is one of the few shark species which is sought after as food in both Europe and the United States. In texture and appearance, porbeagle and mako steaks are both leaner than swordfish and are of a lighter color. The Norwegian porbeagle operations as carried out in the Northwestern Atlantic during the early 1960s, virtually wiped out that shark resource which has not since recovered.

10 Logs were provided to each vessel by Bartlett and turned back to him for later analysis by Woods Hole and Bureau scientists.

11 By this, Senator Saltonstall meant Canadian imports arriving in the United States by truck or freight vessel, as no Canadian fishing boats were actually unloading in United States ports.

12 He was retrieved and lived.

13 According to the FDA, landed swordfish were only legal commodities if the fish were caught within three miles of the shore of Massachusetts where the fish were to be off-loaded *and* if they were then consumed within that state. To the first criterion, no swordfish were probably ever taken within three miles of Massachusetts during this time frame; nevertheless, for the FDA to prove otherwise was practically impossible. An increasingly large percentage of the fish landed in Massachusetts were being shipped in from across state lines; but it also proved difficult for the FDA sleuths to prove, much less apprehend the violators.

14 One American fishing vessel headquartered in the early 1970s out of Woods Hole, MA, made a steady business of meeting Canadians at sea and purchasing swordfish for subsequent resale in Massachusetts. The operator of the American vessel equipped his boat with longline gear; but the gear (from the author's personal observations) showed no sign of wear as would have been the case had it ever been fished. Smuggling of other commodities was carried out by an unknown number of swordfish boats during the 1980s. In 1982, the longliner *Tiki X* was apprehended at Fairhaven, MA, with 30 tons of marijuana

aboard. The following year, the 75-foot *Xiphias* was caught with 27,500 pounds of the same. Charles M. Fuss, Jr. *Sea of Grass.* Annapolis, MD: Naval Institute Press, 1996.

15 Fish running below 75 lbs. dressed weight.

16 In the interest of privacy for the principles involved, monetary amounts contained in the original documents have been deleted by the author.

17 By the time the *Chilmark Sword* was completed, extras had considerably increased the capitalization figure given in this projection.

18 The *Green Sheet* was a daily report issued by the National Marine Fisheries Service (and its predecessors). Its purpose: the listing of poundage and prices reported by the wholesalers (commission merchants) as being average prices daily received from retailers, small secondary wholesalers, restaurants, etc.

19 This seems to have been an inordinately large incidental catch of sailfish. Sailfish as a by-catch of swordfish longline sets was nowhere near as large as this in recorded American catches -- at least those made off the U.S. east coast and Gulf of Mexico. It is likely that the "sporting vigilantes" widely exaggerated their find for the shock value that the article offered.

20 *The Company* collected on the life insurance it had taken out on Bjarne in 1964.

21 A remnant of the cold water mass which made up the Labrador Current to enter the New York bight continues to the southward. Its set runs close to the shores of New Jersey and the Delmarva Peninsula. Its volume increasingly lessens, and its width becomes constricted against the shoals of the outer Cape Hatteras banks by the greater volume of the north-setting Gulf Stream as that system passes close to Cape Hatteras. After passing Cape Hatteras, the cold mass continues south but decreasing in velocity, to widen again as it slowly sets down the coasts of the Carolinas, Georgia, and northern Florida. By the time it reaches Cape Canaveral, it becomes hardly noticeable in velocity, its temperature effect disappearing entirely upon reaching the latitude of Stuart, Florida.

22 The one counter to this concept was the fact that an active winter billfish fishery off the northern coast of Cuba included the broadbill swordfish.

---

[23] There are a number of other locations within the Caribbean and the Gulf of Mexico where spawning also takes place, but in level of importance, they appear to be secondary to the overall dynamics of the swordfish population. From these other spawning locations, which are distributed over an area from the West Indies westward throughout the entire span of the Caribbean and the Gulf of Mexico, it was established that swordfish were present and that they spawn there during the winter months. Within the Gulf of Mexico, the greatest part of the spawning seems to take place in the Bay of Campeche. Some very young larvae have also been collected in the north-central Gulf of Mexico southward of the Mississippi delta.

[24] *Barracudinas* are migratory, traveling in schools, and are generally encountered over the outer slopes in deep water. The form of *Barracudinas* most common in northern waters of the western Atlantic is the white barracudina which is a deep water fish found between 100 and 250 meters, being part of the scattering layer.

[25] The Japanese and Taiwanese longline fishery for tuna during daylight hours is conducted at various mid-water depths, often matching that of the scattering layer. This is a daytime fishery. It would accordingly follow that if swordfish feed during the daytime at mid-water levels, then a heavier incidental catch of swordfish would occur in conjunction with the tuna fishery. But such is not the case.

[26] The author's own observations, from examining stomach contents of reef-inhabiting red snapper *(Lutjanidae)* and grouper (*Mycteroperca,* and E*pinephelus*) caught commercially off the east coast of Florida, is that "parrot beaked" fish are generally restricted to waters within the 70 meter curve. This would seem to indicate that swordfish may at times forage well up on the shelf in Florida latitudes. The bottom fauna there is far richer upon the coral banks (generally located in waters between 55 and 120 meters) where such forage fish congregate. The ranging by swordfish into waters of 70 meters or less therefore seems irrefutable in latitudes south of Cape Hatteras.

[27] J. S. Briggs, *A List of Florida Fishes and Their Distribution,* Bulletin, Florida State Museum, Biol. Sciences, 2 (8): 233-318, 1958. A. H. Leim and W. B. Scott, *Fishes of the Atlantic Coast of Canada,* Bulletin, Fishery Research Board, Canada, No. 155, 1966. Compared to Florida, Canada's nutrient-rich continental shelf produced during the 1970s nearly a thousand-fold more in the amount of fish tonnage of marketable varieties when judged in terms of market receipts.

---

[28] "Dovekie" as described by Roger Tory Peterson in his *A Field Guide to the Birds -- Eastern Land and Water Birds* is

> *"about the size of a starling (71/2 - 9"); small, chubby and neckless...black above and white below, together with its small size and very short stubby bill, render it quite unlike anything else. It is by far the smallest of the wintering sea birds.... It is of the* Alcidae *family which is composed of duck-like sea birds which have a buzzy flight and straddle-legged look when about to land."* The photograph showing these birds after their removal from the two swordfish has been reproduced in this volume.

[29] These fish *(Barracudinas)* have been observed (apparently from submersibles) to swim vertically in a snout up or snout down position. Scott and Tibbo felt that this is a behavioral characteristic that would make them particularly susceptible to the slashing attacks of a swordfish.

[30] Steve Lowe of Fort Pierce, Florida, related this information to this author in 1978: Mr. Lowe was a pioneer of the deep water tile fishery off eastern Florida in the early 1970s. That fishery was conducted from 30'-50' boats dragging bottom drogues to slow their drift along the 200-meter curve at the edge of the Gulf Stream. The tile fishery was carried out with power reels and wire lines mounting two or more hooks per line. These tile fishermen appear to have made the first commercial observations of *Xiphias gladius* along the east coast of Florida (north of Miami). Except in one instance, members of this fleet made no concentrated attempt to catch swordfish until the fishery in that area was exploited by northern longliners beginning in the spring of 1976.

[31] It has been suggested by some fishermen that a proportion of such "bite offs" may be caused by the great white *(Carcharodon carcharias)*. This of course is quite possible; however, the shark catch records as compiled by the Borden Milk Company in the 1940s and 1950s based out of Salerno, Florida, wherein heavy chain and cable gear was used, did not record many great whites. (Source: Conversation in 1976 between Gibson and Jack Casey. During the 1970s, Casey was Chief, Oceanic Gamefish Investigation, while at the National Marine Fisheries Service's Northeast Fisheries Center.) Borden Milk Company had developed the shark fishery out of Salerno as a source of Vitamin A which was extracted from the livers in the days before Vitamin A was synthesized.

[32] Most sharks must free-swim in order to breathe. The restricted tether and spacing of the longline branch lines used in the 160s and 1970s did not allow the majority of sharks sufficient room for the movement they require; thus, when on the longline, many shark species died from suffocation. The

---

ability of sharks to withstand fishing pressures is far lower than that of the swordfish. The shark, unlike the egg bearing swordfish, delivers its young live and in relatively small numbers.

33 With the advent of the universal use of monofilament gangions in the longline fishery beginning in the late 1970s; the ratio of dead sharks as against live sharks caught on longline gear lessened. This was thought to have been the result of much longer gangions which allowed more freedom of movement of the hooked fish and more bite-offs.

34 The sea lamprey is eel-like in appearance. It attaches itself to the host fish by a cup-like suction appendage located on the lamprey's head. Through this, it sucks the fish's blood. On smaller fish, this can result in the death of the host.

35 The jellied condition does not reportedly appear to be encountered in the market place of the 1990s. During the days of the harpoon fishery and the early longline fishery, fish were often landed within only a very few days after being captured. During the 1990s -- except for swordfish taken in the Florida fishery -- most fish have been taken from relatively distant grounds. Since the jellied condition normally makes its appearance shortly after a fish is caught -- usually within three or four days -- the condition is now probably spotted before a fish enters the marketplace and therefore affected fish are discarded. One Maine retailer who has handled swordfish year-round starting in the 1980s informed this author that he has never seen a 'jellied' swordfish.

BIBLIOGRAPHY

"Tuna and Swordfish Distribution studies in Western North Atlantic." *Commercial Fisheries Review* 26(9): 28, 19-30, 31. Fig.

Anonymous. "Larval specimens of tuna, swordfish, and marlin collected in the tropical Atlantic." *Commercial Fisheries Review.* 26 (11):48-50, 1964.

Bartlett, Martin. Field notes and other files, including voyage reports, loaned to the author by Mr. Bartlett. Bartlett was a fishery specialist employed during the 1960s by Woods Hole Institute of Oceanography and later by the National Marine Fisheries Center at Pascagoula, Mississippi, to be followed by his employment by the State of Georgia's Marine Extension Service. During the 1970s and 1980s, Bartlett owned and operated the *Penobscot Gulf,* a swordfish longline vessel which he fished from the Gulf of Mexico north to the Canadian banks.

Beckett, J. S. *Swordfish, Shark, and Tuna Tagging, 1961-69.* Fisheries Resource Board Can., Technical Report 193, 13 p. 1970.

Berkeley, S. A. and D. P. deSylva, *Fishery and Biology of Swordfish in Southeast Florida.* Berkeley and deSylva were the principle investigators for this Sea Grant Project. Miami, FL: University of Miami. Filed with NOAA, project No. R/FR-13, July 1978.

Bigelow, H. B. and W. C. Schroeder. *Fishes of the Gulf of Maine.* U. S. Fish and Wildlife Service Fisheries Bulletin 74, Vol. 53, pp 1-577. 1953.

Blue Water Fishing: Tuna, Mahi-Mahi, Shark and Swordfish. NFI Information Series, Vol. I, Issue 5. Arlington, VA: National Fisheries Institute.

Bremer, Jaime R. Alvarado; Jaime Mejuto; Thomas R. Greig; Bert Ely. "Global Population Structure of the Swordfish (Xiphias gladius) as Revealed by Analysis of the Mitochondrial DNA Control Region," *Journal of Experimental Marine Biology and Ecology,* 1996.

Briggs, J. S. *A List of Florida Fishes and Their Distribution*, Bulletin, Florida State Museum, Biol. Sciences, 2 (8): 233-318, 1958.

Swordfish and (Their Attacks on) Ships. Field, London 57: 663. 1881.

Caddy, J. F. *A Review of Some Factors Relevant to Management of Swordfish Fisheries in the Northwest Atlantic.* Department of the Environmental Fisheries and Marine Service Research and Development Directorate. Technical Report No. 633. 1976.

Carey, Francis G. and Bruce H. Robison. "Daily Patterns in the Activities of Swordfish, *Xiphias gladius*, Observed by Acoustic Telemetry." *Fishery Bulletin,* Vol. 79, No. 2. 1981.

Christie, Donna R. "The Georges Bank/ Gulf of Maine Boundary Dispute Between the United States and Canada," within Georges Bank. Cambridge, MA: MIT Press, 1987, pp 469-73.

Cramer, Jean, *Large Pelagic Logbook Newsletter.* Miami, FL: National Marine Fisheries Service, 1993 and 1995.

Forté, Donald, Chilmark Fleet Corporation's corporate correspondence; Internal Revenue Service tax returns for years 1964-69.

Gibson, Charles Dana. "Victim or Participant? Allied Fishing Fleets and U-boat Attacks in World War I and II," Northern Mariner. Ottawa, Canada: Canadian Nautical Research Society, Vol. I, 4, October 1991.

Goode, G. Brown. *The Fisheries and Fishing Industries of the United States.* Volume I, Section V, History and Methods of the Fisheries, "The Swordfish Fishery." Washington, DC: Government Printing Office, 1887.

Gudger, Eugene W. *Fishery Biology of the Swordfish (*Xiphias gladius Linnaeus Teleostomi: Xiphiidae) *in Cuban waters.* Fisheries Research Board of Canada. Trails. Seer. 611. 1965.

Gudger, Eugene W. *The Alleged Pugnacity of the Swordfish and the Spearfishes Shown by Their Attacks on Vessels* (a study of their behavior and structures which make possible their attacks). Mem. Roy. Asiatic Soc., Bengal 12(2): pp 215-315, 1940.

Hare, G. M. *Atlas of the Major Atlantic Coast Fish and Invertebrate Resources Adjacent to the Canada- United States Boundary Areas.* Fisheries and Marine Services, Technical Report 681. St. Andrews, New Brunswick, Canada. 1977.

Hoey, John J. and John G. Casey, unpublished report entitled "The United States Swordfish Fishery, 1963 - 1982: Distribution and Trends in Catch and Effort for the New England Fleet." Narragansett, RI: National Marine Fisheries Service, Northeast Fisheries Service, 183.

Hoey, John J. and John G. Casey. Draft: "The U.S. Swordfish Fishery, 1963-1982: Distribution and Trends in Catch and Effort for the New England Fleet." Northeast Fisheries Center, Narragansett, RI. 1983.

Hoey, John J.; Ramon Conser; Angelo R. Bertolino, "The Western North Atlantic Swordfish" in *Audubon Wildlife Report, 1989 - 1990,* edited by William Chander. San Diego and New York: Academic Press, Inc., 1990.

ICCAT -SCRS 94/123. Hoey, John J. A. Bertolino; Christopher W. Rogers, 1994.

ICCAT Working Document, SCRS 95/74, National Report of the United States: 1995. NOAA, National Marine Fisheries Service, Miami, FL. Contribution MIA-95/96-03.

ICCAT-SCR 94/122, Hoey, John J.; Jaime Mejuto; Julie M. Porter; Heath H. Stone; Yuji Uozumi, *An Updated Biomass Index of Abundance of North Atlantic Swordfish, 1963 - 1993.* Arlington, VA: National Fisheries Institute, 1994.

Ingle, R. M. *Summary of Florida commercial marine landings 1970.* 64 p illus. Map U.S. Fish and Wildlife Service, Bureau of Commercial Fisheries: Washington, DC. 1970.

International Game Fish Association Yearbook, 1952. World summary by months, geographical sections, and species as to where, when and under what conditions various marine game fishes may be caught. Includes marlins, sailfish, and swordfish.

Kingsley, J. S. "The food habits of the swordfish." *Science,* 56: 225-226. 1922.

Kramer, David. "A record of the swordfish Xiphias gladius Linnaeus from the Texas coast". *Copeia* 1950 (1): 65.

LaMonte, F. "Notes on Breeding Grounds of Blue Marlin and Swordfish off Cuba. *Copeia*: 258. 1944

Lauzier, L. M. *Larval Swordfish* (Xiphias gladius) *for Three Localities in the Western Atlantic.* Fisheries Research Board of Canada. 26(12): 3248-3251. 1969.

Lee, R. E. "The Occurrence of Female Swordfish in Southern New England Waters, With Description of Their Reproductive Condition. *Copeia, pp* 117-119. 1942.

Leim, A. H., and W. B. Scott. *Fishes of the Atlantic Coast of Canada,* Bulletin, Fish Resources Board, Canada, No. 155, 1966.

M'Gonigle, R. H., and A. H. Leim. *Jellied Swordfish.* Biology Board of Canada, Atlantic Prog. Report No. 19, pp 3-5. 1937.

Markle, G. E. *Distribution of larval swordfish in the Northwest Atlantic Ocean.* NOAA Technical Report, NMFS SSRF 675: pp 252-260. 1974

Nakamura, Izumi. *Some Aspects of the Systematics and Distribution of Billfishes.* Proceedings of the International Billfish Symposium. Kailua-Kona, Hawaii, 9-12 August 1972. Part 2, Review and Contributed Papers. NOAA Technical Report NMFS SSRF - 675. 1974

Nigrelli, Ross F. "Parasites of the Swordfish, *Xiphias gladius Linnaeus.*" *American Museum Novitates* 996: pp 1-16. 1938.

Rich, W. H. "The Swordfish and the Swordfishery of New England." *Proceedings of the Portland Historical Society of Natural History.* 4(2): pp 1-102. 1947.

Richards, William J. "Evaluation of identification methods for young Billfishes." *Proceedings of the International Billfish Symposium,* (Kailua-Kona, Hawaii, 9-12 August 1972). Part 2, Review and Contributed Papers edited by Shomura, R. S. and F. Williams. NOAA Technical Report NMFS SSRF - 675, pp 62-72. 1974.

Scott, W. B. and S. N. Tibbo. *Food and feeding habits of swordfish, Xiphias gladius, in the western North Atlantic.* Journal Fisheries Res. Board Canada, 25 (5): pp 903-919. 1968

Scott, W. B. and S. N. Tibbo. *Food and feeding habits of swordfish, Xiphias gladius, in the Northwest Atlantic Ocean.* NOAA Technical Report NMFS SSRF 675. Pp 138-141. 1974.

SCRS 90/33, Hoey, John J. and Jaime Mejuto. Swordfish Size Composition Data From Spanish and United States North Atlantic Fisheries. Arlington, VA: National Fisheries Institute, 1990.

SCRS-90/93. Mejuto, Jaime and John J. Hoey. *An Approach to a Stock Hypothesis for the Swordfish (Xiphias gladius) of the Atlantic Ocean.* Arlington, VA: National Fisheries Institute, 1990.

Springer, Stewart and Harvey R. Bullis, Jr. *Collections by the Oregon in the Gulf of Mexico.* US Fisheries and Wildlife Service Special Science Report. Fish, 196, p 72. 1956.

Squire, James L., Jr. "Marlin and swordfish in oceanic waters of the western North Atlantic." *Copeia* 1962 (1), pp 216-219. 1962.

Tibbo, S. N. and L. M. Lauzier. *On the origin and distribution of larval swordfish (Xiphias gladius L.) in the western Atlantic.* Fisheries Res. Board Canada Technical Report 136. 1969.

Tibbo, S. N., L. M. Lauzier, and A. C. Kohler. "Canadian fisheries research extends to Caribbean Sea." *Trade News,* Department of Fisheries, Canada. 17 (11-12), pp 3-9. 1965.

Tibbo, S. N., L. R. Day, and W. F. Doucet. *The Swordfish (*Xiphias gladius L.*) Its Life-history and Economic Importance in the Northwest Atlantic.* Fisheries Res. Board Canada, Bulletin 130, pp 1-47, 1961.

Ueyanagi, Shoji, Y. Nishikawa, Kikawa, and Uto. *Distribution, Spawning, and Relative Abundance of Billfishes in the Atlantic Ocean.* Far Seas Research Laboratory, Shimizu 424, Japan. 1970.

Wathne, Fredrick. "Summary Report of Exploratory Longline Fishing for Tuna in Gulf of Mexico and Caribbean Sea, 1954-1957." *Commercial Fisheries Review.* 21 (4), pp 1-26. 1959.

Wilson, Peter C. and Martin R. Bartlett. *Inventory of U.S. Exploratory Longline Fishing Effort and Catch Rates for Tunas and Swordfish in the Northwestern Atlantic, 1957-65.* U.S. Department of Interior, Fisheries and Wildlife Service, Bureau of Commercial Fisheries, Special Science Report -- Fish. No. 543. 1964.

Wise, J. P. and J. C. LeGuen. "The Japanese Atlantic Longline fishery, 1956-1963." *Proceedings of the Symposium on Oceanography and fish.* Resources of the Tropical Atlantic -- Review papers and contributions. UNESCO. Paris, pp 317-347. 1969.

Wise, J. P., and C. W. Davis. *Seasonal Distribution of Tunas and Billfishes in the Atlantic.* NOAA Technical Report. NMFS SSRF 662. pp 1-24. 1973.

Yabe, H. "Larvae of the Swordfish, *Xiphias gladius.*" Japanese Journal *Ichyol.* L, pp 260-263. 1951.

INDEX

About the Author

Charles Dana Gibson is the author of six books on maritime and naval history. His last two books, co-authored with his wife, received the 1996 John Lyman Book Award for U.S. naval history. Gibson's association with the commercial swordfishery ocurred during the 1970s when he owned and operated *John de Wolf II,* a longline vessel based out of New Bedford, Massachusetts. (See photo below.) He fished offshore grounds stretching from the Canadian banks to the Straits of Florida. Gibson is presently working on a forthcoming volume of The Army's Navy Series (Ensign Press) which will examine the U.S. Army's transportation effort by sea, 1868 - 1950. Previous volumes cover the periods 1775 to 1860 and 1860 - 1868.